REDISCOVERING
GOLD

IN THE
21ST CENTURY

by

Craig R. Smith

Foreword by
Pat Boone

Idea Factory Press

The cover design was inspired by the Lewis & Clark Expedition (1804-1805) who were among the brave trailblazers of American history. The coins pictured are the obverse and reverse of the Lewis & Clark $1 Gold Commemoratives minted in 1904 & 1905 to commemorate the 100-year anniversary of the expedition. These golden gems were the highest performing gold coins in 1999-2000, and offer all investors a peek at the first gold rush of the new millennium.

Executive editor and cover design: David Bradshaw
Contributing writers: Earl Brown, Dr. Fred Goldstein
Copy editor: Beth Cullum
Illustrations: Cliff Vaughn

PRINTED IN THE UNITED STATES OF AMERICA
First Edition: July 2001 - Second Edition: May 2002 -
Third Edition: February 2003 - Fourth Edition: March 2004-
Fifth Edition: February 2005

ACKNOWLEDGEMENTS

First I would like to acknowledge my pastor Tommy Barnett for his encouragement to write a book and his constant support and guidance to always do the right thing. He is my hero and mentor - without him I'm not sure I would be alive today. Thanks also to my trusted associate and friend, Steven Carnow and the entire management team at Swiss America for their faithfulness over these past 20 years. I'm eternally grateful for the love and strength of my wife Melissa who has always stood by my side during the difficult times encouraging me, regardless of how dark the storms looked. I dedicate this book to my daughters, Holly and Katie, who are my legacy. You have been wonderful children and a great blessing in our lives. Special thanks to David Bradshaw of My Idea Factory, who has helped me to communicate my heart and mind over the last two decades. Above all, this book is a labor of conscience, and for that I am thankful to God. - *Craig R. Smith*

CONTENTS

FOREWORD

By Pat Boone

You hold in your hand a simple road map to rediscover gold in the 21st century. For years I've followed Craig Smith's financial road map carefully and my tangible portfolio has grown appreciably.

Wise management of money is one of the greatest needs of our times. Collecting and investing in United States coins offers parents and grandparents a great educational tool - and above-average returns.

In 2001 the American economy began slowing down. Some economists were forecasting a recession or worse, causing many businesses to cut back. But not Craig Smith - he is expanding in 2001!

Craig Smith reminds me of Joseph - a humble yet tenacious man who was given divine understanding of an approaching national crisis. Joseph organized his nation to prepare by storing tangible resources, and during one of the worst economic famines in history, provided the needed grain to feed his nation... and save his loved ones as well.

Craig brings the reader good news about how to hedge your family against any financial circumstance. His number one financial commandment is, "Thou Shalt Diversify Assets." Follow it!

Ready, set, go for the gold!

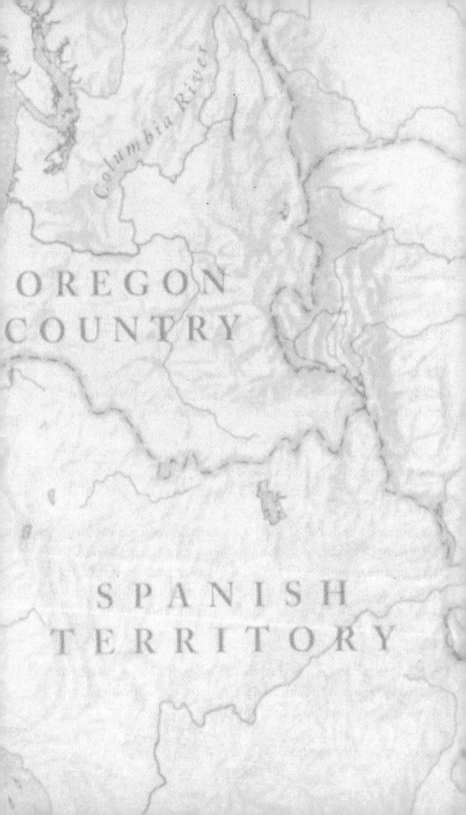

PREFACE

LET'S REDISCOVER GOLD!

Having spent the last 30 years participating in the gold market, you could say I've earned the right to speak out on the subject.
And I often do, as president of Swiss America Trading Corporation.

During radio and television interviews I'm often asked whether I think that gold will go up, down or sideways. My answer? "Yes, it will go up, down or sideways - depending on the type of gold you own."

But the real questions about gold ownership should be: "Which type of gold offers the best financial protection? The most tax advantages? The most privacy? The maximum profit potential?

If you've ever purchased gold coins, this book is written for you. If you've never invested in gold coins, this book is written to challenge you to consider gold ownership in the 21st century.

Few investors have yet to discover the value and safety of having a truly diversified portfolio. Most simply follow the crowd, evidenced by the fact that last year most Americans positioned more than 60 percent of their money in the equities market - a historic high.

Speculators in today's financial and stock markets have become very demanding (read: spoiled) over the last decade. They want it all - value, growth and peace of mind. But the search for a "perfect investment" is elusive and usually ends in disappointment.

The lack of asset diversification is also true when it comes to gold. For example, three years ago based upon ominous news stories, many feared the year 2000 computer bug might bring on a global financial crisis and decided to buy gold bullion coins, just in case.

Many bullion owners have told me that they feel like they've been punished for preparing for the worst because their bullion coins - like American Eagles, Krugerrands, Maple Leafs and British Sovereigns - have dropped in price since 1999.

Adding insult to injury, most financial pundits have now declared that gold has finally outlived its monetary role forever and is merely the fading icon of the old economy. The mass media has nearly convinced Americans that gold is no longer a viable store of value or a needed reserve for the U.S. dollar or other currencies.

But bullion owners aren't the only ones disillusioned in 2001. Millions of Americans who rushed into buying dot-com stocks - which dropped 60, 70, 80 percent or more since their market peak - are now asking, "Should I buy, sell or hold?"

Now that the tech binge is over, it's time to sober up to some new realities. I want to introduce a strategy that will help you sleep soundly even if every tech or Internet stock you own drops to zero.

Imagine discovering a special type of gold that taps into the positive economic trends as we enter the 21st century - then you'll understand my excitement... like finding a treasure map with a big "X."

You see, not all forms of gold have lost value during the last few years. Certain U.S. gold coins have grown more than 100 percent during the same period that some tech stocks shrunk 80 percent.

My strategy of gold ownership offers something for every investor: a balanced portfolio that will offer protection against a financial disaster as well as realistic, sustainable profit potential.

Now that I have your attention, let's begin with the age-old question "Why gold?" and then reflect on the monetary and investment role of gold throughout American history. "If we don't learn from history... we are destined to repeat it," as the saying goes.

1

A Short History of Gold

1.1 Why Gold?

SINCE the dawn of creation, gold has been held in high esteem as a store of value and a universal monetary substance in all civilizations. But why?

Pleasant to the touch, highly malleable, resistant to all external influences, deep yellow in color and with deep luster, gold has captivated mankind and won a special place in human history. Clearly, this love of gold is not limited to the capitalist economic system.

Gold was greatly sought after in Greek and Roman times and throughout the Middle Ages. It is prized in the Middle East where it was used for the first time as money, and it's also highly valued in China, Africa, the pre-Columbian America of the Aztecs and Incas, in the Russian Empire and even among the followers of Attila and Genghis Khan.

Thus gold is one of the few common values that has united mankind throughout the millennia, transcending race, religion and geography - a rarely noted fact but significant in light of today's growing cultural convergence and emerging global economy.

After thousands of years serving as the world's money in the form of coinage, the advent of paper and electronic money is now challenging gold's monetary role in the 21st century.

Gradually, throughout the last several decades, gold has assumed a new role as a monetary reserve medium, rather than a medium of exchange. This subtle yet powerful monetary shift has gone mostly unnoticed by the American public.

By 1999, the powerful central banking community that controls all modern money creation began advocating the complete abandonment of gold as a monetary reserve.

Is this historic shift the result of the natural progression of human maturity? Is this the natural evolution of money, as Darwinian - Keynsian economists would say?

Or, is it a quantum leap toward total manipulation of the masses into a single manageable global political, social and economic system?

These questions are growing in the minds of many Americans. Let's quickly review America's monetary history and see if we can find some answers.

1.2 Substance Over Symbolism
The Folding, Spindling & Mutilating of America's Money System

Imagine for a moment that you have the ability to create any amount of money, without ever having to produce anything. Is there anyone or anything you couldn't buy? Probably not.

Sound impossible? It should be, but it isn't. Just ask your local Federal Reserve banker - they do it every day.

The folding, spindling and mutilating of America's monetary system became legitimized in 1913, when the Federal Reserve was formed. Long ago bankers discovered a nasty little secret referred to as "fractional-reserve banking" which is fueled by credit and debt creation out of thin air.

The modern American monetary system is the result of an incestuous relationship between the federal government and the private banking cartel, deceptively called The Federal Reserve System (a.k.a. "The Fed").

But don't expect the mainstream press or prominent political figures to ever discuss this relationship publicly. Sadly, few Americans understand the process, or even challenge the Fed's attempt to manipulate the money system.

In the two centuries prior to the creation of the Fed, unredeemable paper currencies were judged as unethical and immoral. As of 1792, they were deemed unconstitutional as well.

The fundamental misconception today is that America's paper or electronic currency, denominated in Federal Reserve Notes, is that a dollar actually has any intrinsic value.

In the words of former Fed economist John Exter, "Today's U.S. dollar is nothing more than an IOU-nothing." Paper money retains only the symbol, or form, of its original substance - gold and silver.

Let's now examine the untold story of how and why the U.S. dollar was transformed from substance (gold) to symbolism (debt) - and what you can do to recover the substance while you still have time.

Legal Plunder

As difficult as it is for honest, hard-working Americans to fathom, the lifeblood of the American political and economic system is legal plunder. The 19th-century economist Frederic Bastiat summed up the tendency of central governments to embrace economic plunder in this way:

> **"There are two ways to acquire the niceties of life: to produce them or to plunder them. When plunder becomes a way of life for a group of men living together in society, they create for themselves in the course of time, a legal system that authorizes it and a moral code that glorifies it."**

The gradual devaluation of U.S. currency during the 20th century reflects a more subtle transformation - too many Americans have abandoned the morality and economics of our Founding Fathers.

Today's warped and degenerate political system represents a marked departure from the statesmanship of a bygone era. Economics likewise has degenerated into a convoluted science orchestrated to conceal a colossal fraud perpetrated on an unsuspecting public.

In short, "We the People" allowed the Federal Reserve, with the full cooperation of the federal government, to replace the "Puritan work ethic" with a "pagan plunder plan" and now the chickens are starting to flock home to roost.

To achieve this massive wealth distribution plan required a shift in public values from hard work and responsibility, to hardly working and gambling. This dramatic change has occurred gradually over the past two or three generations.

The result or fruit of this shift can be seen in the monetary realm. We abandoned true money (commodity - gold or silver) in favor of false money (fiat - paper, electronic). Here's how it happened, in a nutshell.

Substance Money

All true money must be derived from a commodity, or at least have a substance to back it up, or it will gradually become fraudulent, or fiat money.

Historically, the most common substance used as a medium of exchange and a store of value has been gold or silver coins of a standard weight and fineness.

THE U.S. COINAGE ACT OF 1792 specifically defined a dollar as "one twentieth of an ounce of gold (25.8 grains of 90 percent fine) or a silver coin containing one ounce of silver (421.5 grains of 90 percent fine)." The Founding Fathers specifically prohibited the federal government from issuing Bills of Credit, (paper money) in the U.S. Constitution.

Congress shall have Power to coin money and regulate the value thereof ... No State shall make any Thing but gold and silver Coin a Tender in Payment of Debts.

-Art.1 Sec. 8 & 10

America's system of constitutional, commodity-based money functioned well in our nation for 125 years, from 1792 to 1913. Then "We the People" made a big mistake - we allowed a privately owned corporation called the Federal Reserve to begin creating *paper* money instead of gold and silver coins as the Constitution requires.

Trust Money

The Federal Reserve's monetary manipulation began with a promise to create paper money that could always be redeemed for commodity money - gold or silver coin. This 100 percent redeemable money is referred to as *fiduciary* or trust money.

The creation of fiduciary money assumes that the promise of payment in substance by the issuer is redeemable at some future point. Trust money was used as a medium of exchange even though it consisted largely of an intrinsically valueless substance - paper.

Since the U.S. government was prohibited by constitutional law from issuing this trust money, the Fed - a private corporation - was created to soften and manipulate the economic down-cycles in 1913. The price we have paid is surrendering our substance money (gold) for trust money (credit/debt). In my view, central bankers took the mine... and we got the shaft. Why do I say that?

History has proven time and again that neither bankers nor governments possess the discipline needed to limit the amount of credit (or paper money) to equal the true supply of gold and silver coins. So the supply of paper money (credit/debt) must continually rise.

The result is always disastrous in the long term because the economy suffers through cycles of inflation, deflation, artificial growth, recession and depression. Because U.S. citizens did not protest the use of trust money, our economic system then began to degenerate into untrustworthy or *fiat* money.

Fiat Money

Fiat paper money abandons any promise whatsoever to redeem the paper currency in any physical commodity. This third step in the decline of our currency is considered by many historians and economists as the beginning of the end, monetarily.

Dr. Franz Pick, the noted Austrian economist, aptly stated the link between a nation and its money,

> **"The destiny of a currency is, and always will be, the destiny of a nation."**

Under the fractional-reserve banking rules, a bank must always issue more units of fiat money than can ever be redeemed (typically at an 8:1 ratio). Fractional-reserve banking is inherently a fraudulent system. But by 1933 FDR forced Americans off the gold standard and onto the treadmill of credit fueled by fractional-reserve banking.

Here is an example that may help you grasp why fractional banking is flawed. Imagine that you live on a small island with just one other inhabitant - a fractional-reserve banker. On the island there is only $1,000 in circulation total. Let's say you decide that you want to start a fishing business, and visit your banker for a loan. The banker agrees to loan you $1,000 but must charge you $50 interest. That means you will owe $1050. But wait, there is only $1,000 in circulation. Where will the other $50 come from? It must be created by the banker or you could never fully pay the loan back. This is the origin of inflation and devalues every other dollar in circulation.

Therein lies the faulty foundation of fractional banking - it must constantly inflate the amount of currency which in turn decreases the value of all the money in circulation. The consequences are many, but most harmful is the crushing of the middle class via long-term monetary inflation. If (when) the public finally discerns that the Emperor (Fed) has no clothes, I expect hyper-inflation and a flight back to substance money in a New York second.

Lenin pondered this modern flaw in the Capitalist system stating,

> **"The best way to destroy the Capitalist system is to debauch the currency. The process of inflation is so insidious that not one in a million can properly diagnose it, until it is too late."**

The Federal Reserve and the federal government are banking on Lenin's conclusion - that the public will not become aware of this insidious process… until it is too late.

Karl Marx also knew that centralized money control was critical to control the masses.

> **"Centralization of credit in the hands of the State, by means of a national bank with State capital and an exclusive monopoly."**
> -5th Plank, Communist Manifesto,
> by Karl Marx (1848)

The facts are that popular delusion and public confidence are the only two forces that uphold our present fiat money system. The financial house of cards created by our massive personal, corporate and government debt is now more vulnerable than ever. The government knows it and the Fed knows it - they even admit it in print:

> **"All the paper money issued today is Federal Reserve Notes. The real backing for the nation's money is faith in the strength, soundness and stability of the U.S. economy."**
> -*Hats the Fed. Reserve Wears*, Federal Reserve Bank of Phil., p. 4

One can only deduce that the Fed believes that as America's faith and confidence goes, so goes the economy. It is interesting that the root meaning of the word credit (*credaria*) is "to believe." It is true, we now have a monetary system purely based on faith - faith in a system that betrays us and our children.

Purchasing Power of the Dollar
(1792 = 1.00)

The solid portions of the curve show periods when the dollar was redeemable into monetary commodities (gold or silver) and the broken portions are periods when redeemability at fixed rates was impaired. The circled portions show periods of disinflation or deflation.

Note: Purchasing power was calculated from the Wholesale Price Index
(source: US Department of Labor)

Over the last decade the government has even established a formal "Consumer Confidence Index" as a means of monitoring and manipulating the public confidence in the economy and money system. This index has been moving downward in 2001, reflecting a loss of confidence in the Fed... and our money system.

Virtual Money

Starting in 1990, Federal Reserve Notes have two subtle additions: a metallic strip embedded in the bill and special micro-print around the President's bust. The official reason is to thwart counterfeiters and monitor anyone attempting to leave the U.S. with a suitcase full of cash through the use of special airport detectors. This sounds reasonable enough, right? After all, counterfeit Federal Reserve notes are popping up all over the world. They're calling it *economic terrorism*. The Fed cannot allow competition in the money counterfeiting business to encroach on its domestic policy of issuing unconstitutional fiat (read: counterfeit) money.

The next major step is to convince Americans to convert entirely to a totally intangible, electronic money system - with no "cash" at all.

It took the last decade to prepare us, but I expect over the next decade the government and banks will accomplish it because it offers convenience - the new passion of American culture.

So, where do we go from fiat money? To what I call virtual money - that is, pure credit transactions reduced to blips on computer screens. This new form of money gives the government total economic control over the populace - a goal many have long desired. Financial privacy is also forfeited in the process.

Should all of this give you grave concern, or a sleepy nod? It depends on how much you value your privacy, sovereignty, freedom, liberty and that of the next generation - all of which are God-given rights under the U.S. Constitution.

The steady decline in the value of buying power throughout the past 80 years is a crime in my book. In fact, a 1900-dollar is worth less than 3 cents today due to inflation.

What can be done? How can we recover an honest foundation for economic stability - even if our government won't? One person at a time. The good news is that we still have options and rights.

Coined Freedom

Gold and silver coinage has been used as a medium of exchange and store of value throughout all recorded history. From Abraham in the Old Testament to your great grandfather - they all knew that real money represented true freedom and liberty. They also knew that freedom was not free, it often required waging a battle - which was anything but convenient.

Gold and silver coins represent true economic value because they have integrity by design and content. Prior to 1933, U.S. gold coins were the visible evidence of an honest money system. The denomination and value of the coin corresponded with the weight and fineness of the substance - gold. "A just weight and measure," as the Bible demands.

All of this changed overnight when FDR recalled the gold in 1933, making gold ownership illegal and allowing the Federal Reserve to issue fiduciary money, redeemable only in silver, not gold. This trust money was minted until 1965.

Since then America has functioned on debt, credit and fiat money. "Funny money," as G. Gordon Liddy once told me during a radio interview. The truth is that the systematic crushing of the middle class family due to long-term inflation is anything but funny.

In 1934 the government removed gold from circulation and in 1965 they removed silver. Notice that until 1964, U.S. silver coins still represented the economic mandate of just weights and measures. The amazing thing is how few opposed the Fed, perhaps because Americans still trusted the federal government.

Today we readily accept symbolic money instead of substance money with no thought. Our post-1965 copper/nickel tokens circulated today demonstrate a serious departure from our heritage of honest money and represent a gutted economic ethic.

Did you know that even our "copper" pennies are not even made out of copper anymore? Go ahead, scratch a penny with a nail - nothing but pop-metal. A pure copper penny is worth about 3 cents today. Our money system today is symbolism, pure and simple, without any valuable substance to it. For this reason alone I feel that every American should diversify a portion of their fiat money into real money - gold and silver coins. As an added bonus, many historic U.S. gold and silver coins have maintained an above-average track record since the late 1960s.

The nature of our present economic and monetary environment requires decisive action - if not for ourselves, for our children and their sake. As R.E. McMaster Jr. puts it, this is "no time for slaves."

Today, monetary myth is so widespread that it appears that nothing short of a financial meltdown will rattle Americans enough to face reality. Is that what it will take? I hope not, but I fear so.

Remember 1979? The Carter deficits, double-digit inflation and feverish activity in precious metals? Most of us will never forget that year. Something similar or worse may await America - and the time to plan for it is now.

Prominent free market economists like R.E. McMaster, Jr., John Pugsley, Dr. Edwin Vieira, Lew Rockwell, Dennis Peacocke, Bill Bonner, Richard Russell, Mark Skousen, Frank Venerosa and many others agree that we'll face a monetary and dollar crisis soon - based on the huge debt bubble amassed during the 20th century.

U.S. non-financial corporate debt has reached a staggering 46 percent of GDP - its highest level ever. Consumer installment debt is up to 21.7 percent of disposable income - also the highest level ever.

U.S. Personal Savings Rate

In early 2001 the average household has a credit card balance of $7,200 - an amount that would take 30 years to pay off if you made only the minimum payments. Savings rates recently hit zero.

A New War on an Ancient "Drug"

Debt and credit has become America's drug of choice. "Buy now - pay later" has become the new mantra of the last 40 years, but at a heavy price to our money system, our communities and our families.

The modern disintegration of the family, rising divorce rates and bankruptcy are just a few of the visible casualties that can be traced back to debt and credit abuse.

I suggest we admit that debt is America's greatest drug problem today and follow the steps of any good rehab center - to start living within our means. This may require going "cold turkey" for some.

The only alternative (as with drugs and alcohol) is to gradually increase the dosage to continue the 'debt high.' This is what Alan Greenspan & Co. is banking on to keep the economy from going into a full-blown recession in the near future.

I suggest that you "just say no" to all drugs - especially debt - or else face the possibility of becoming a bankruptcy statistic in the great debt wash-out we are heading for unless we change direction.

Sadly, few are prepared to face this potential debt crisis with a financial house that is built on anything more than paper and electrons.

As for me, I sleep a lot better with a solid foundation of historic U.S. gold and silver coins in my portfolio to provide the capital preservation, liquidity and growth needed for the days ahead.

The next big question is which type of gold will offer the best potential for the future - bullion or rare gold? Discover the differences between them for yourself in Chapter 2, *Gold: Bullion or Rare?*

2

Gold: Bullion vs. Rare

2.1 All Gold is Not Created Equal

Identifying the right type of gold to own is
sort of like identifying a trustworthy doctor, lawyer or
stockbroker… part luck and part skill.

For example, many stumble onto a particular market based on a
referral or compelling media interview. The public is still relatively
trusting of media financial pundits - whether they deserve it or not.

But, by failing to ask the right questions before plunking their hard-
earned money into a new market, many have become victims of
their own haste. My goal is to educate you so that you will be better
equipped to make the right choices.

For nearly two decades I've encouraged my brokers to ask their
prospective clients questions like: "What is the primary reason that
you are buying gold - financial safety, financial privacy or financial
growth?" We encourage our clients to include all three elements in
a diversified gold portfolio.

Will Roger once said, "Buy land, they aren't making it anymore!"
The same is true of collectibles.

My motto in gold investing is safety first - and all gold coins provide
financial safety and security. Financial privacy is becoming another
major concern for investors. As you will soon discover, some forms
of gold protect your privacy much better than other forms.

Lastly, which form of gold has the most profit potential - bullion or
rare? In my view, only after you have these answers and understand
these basics are you ready to discuss purchasing.

Many investors are confused about the wisdom of gold ownership today based on negative commentaries. For example in the summer of 1999, *Business Week, The New York Times* and the *London Financial Times* all spoke negatively about gold bullion. Here's just one example from the *London Financial Times:*

> **"Gold may be heading for a fundamental re-appraisal in which its price is based on its intrinsic worth rather than its supposed monetary value."**

Is this true - is gold losing its investment luster? The answer depends on the type of gold we are discussing - bullion or U.S. rare gold.

Gold Bullion Coins

Historically gold bullion has been a good hedge against inflation serving as a portfolio anchor. But during the last 20 years, as inflation fears have diminished, gold bullion has proven to be a poor investment based on the return, albeit good insurance. In comparison to the stock market, mutual funds or even passbook savings

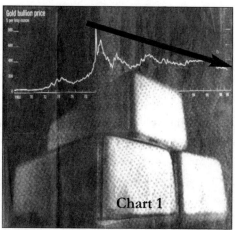

Chart 1

accounts, the only gold bullion rush in the 1990s was to the exits. (See Chart 1)

Contrarian logic says: "Gold is so low that it is a great investment... Buy low, sell high is the best investment strategy in the world." These are true, however from a return-on-investment perspective, gold bullion coins have been a losing investment for most Americans. Buy-and-hold gold bullion purchasers lost at least 20 percent over the last decade. Is that all about to change?

Time will tell. In Chapter six I'll discuss valiant efforts by GATA to expose the alleged financial manipulation of gold prices by central bankers.

If you own bullion, I recommend asking your gold broker questions like: "What was your top product last year, three years ago, five years ago and ten years ago? How have they performed this year?" Demand facts and figures!

So-called "gold dumping" has become a favorite pastime of the international banking elite during the last five years. Central Bank liquidations of gold bullion began almost a decade ago. The Bank of England announced that they were liquidating half of their gold reserves in 1999.

The only short-lived gold rally was in the third quarter of 1999, which was based on Central Bank announcements that they would halt the massive gold selling into the gold futures market. (More on the illegal alleged gold price-fixing in Chapter 6.)

Another big question affecting any gold bullion investor is: Will gold prices rise or fall in the technology-driven 21st century?

Technology & Gold

Technology is making virtually everything in the world less expensive to produce today, including gold mining. Look at the effect of advancing technology on computer prices, they have been dropping 10 to 20 percent each year. Each day it seems you can get more power and memory for less money, yet we have merely scratched the surface of technology.

Gold prices are reduced as more advanced technology processes are used in the exploration and mining of precious metals. Advances in mining techniques have decreased mining costs and increased the supply of gold, thus contributing to the lackluster performance of gold bullion in recent years.

At the time of writing this book gold is trading at $267 an ounce, which is very near the 20-year low and also very near the mining cost. So, by any standard gold bullion is undervalued today and has a much greater upside potential than downside risk.

To understand the difference between how technology affects gold bullion and U.S. rare coins, consider this question: Has technology influenced the price of a rare Picasso painting?

The answer is no, except when you consider that our growing economy has produced more wealth, and wealthy people like to purchase artistic masterpieces. Internet technology has actually had the exact opposite effect - more access to information on the Web tends to drive the prices of rare art and collectibles upward by opening new markets. Just check the latest auction figures from Christie's of New York.

The same is true of U.S. rare coins. Ask yourself: Will technology ever lower the cost of producing a rare U.S. $20 gold coin? No, never, because the U.S. stopped minting them almost seven decades ago. In my view technology can only help to propel rare collectibles of all types to appreciate - especially as demand rises globally via the rapid transfer of information about diverse investment opportunities such as U.S. rare coins on the Internet.

Investment-Grade Gold

The performance of rare U.S. gold coins, also known as invest-ment-grade gold, offers a sharp contrast to gold bullion in recent years.

In fact, by the late 1990s, some bullion dealers even began selling limited edition bullion coins as "rare coins" to compete with the rising rare coin market prices. No wonder uninformed investors are so easily confused.

During the last 20 years as gold bullion prices have drifted lower, many collectible gold coins have established a steady upswing that would make even the equities market jealous. For example, Chart 2 illustrates how the investment-grade $20 Liberty gold coin market moved up between 1995 and 1998.

Chart 2

U.S. rare coin collectors and investors alike enjoy the history, enduring beauty and impressive growth curve of the collectible U.S. coin market.

"Numismatic" (collectible) U.S. rare coins often are passed from generation to generation because they become prized personal possessions.

Many other factors have contributed to the growth of U.S. rare coins, but the most important factor is that collectible rare coins are not minted anymore unlike bullion coins. Rare coins have a shrinking supply and a growing demand.

MS-63/64 $20 Gold Libs Rise 20% to 88%!

GOLD COIN	DATE	GRADE	BUY/'95	VALUE/'97	GROWTH
$20 Liberty	1883-S	MS-63	$4,100-4/95	$4,950-7/97	21%
$20 Liberty	1884-S	MS-63	$2,950-8/95	$4,130-11/97	40%
$20 Liberty	1885-S	MS-63	$2,775-8/95	$4,350-11/97	57%
$20 Liberty	1888-S	MS-63	$2,800-3/95	$3,850-11/97	38%
$20 Liberty	1891-S	MS-63	$1,625-2/95	$2,450-11/97	51%
$20 Liberty	1892-S	MS-63	$1,960-3/95	$2,520-11/97	29%
$20 Liberty	1893-S	MS-63	$1,750-1/95	$2,520-11/97	40%
$20 Liberty	1894-S	MS-63	$1,275-2/95	$2,030-11/97	59%
$20 Liberty	1895-S	MS-63	$1,375-11/95	$1,680-11/97	22%
$20 Liberty	1895	MS-63	$1,025-11/95	$1,260-11/97	23%
$20 Liberty	1896	MS-63	$1,260-3/95	$1,610-11/97	28%
$20 Liberty	1896-S	MS-63	$1,275-11/95	$1820-11/97	43%
$20 Liberty	1897-S	MS-63	$1,120-10/95	$1,365-11/97	22%
$20 Liberty	1897	MS-63	$1,050-11/95	$1,260-7/97	20%
$20 Liberty	1898-S	MS-63	$980-11/95	$1,260-11/97	29%
$20 Liberty	1899-S	MS-63	$1,200-11/95	$1,610-11/97	28%
$20 Liberty	1901-S	MS-63	$1,600-5/96	$2,520-11/97	58%
$20 Liberty	1902	MS-63	$3,775-2/95	$4,650-11/97	23%
$20 Liberty	1902-S	MS-63	$1,960-2/95	$2,660-11/97	36%
$20 Liberty	1903-S	MS-63	$1,150-3/96	$1,470-11/97	28%
$20 Liberty	1905-S	MS-63	$1,750-5/96	$2,170-11/97	20%
$20 Liberty	1906-D	MS-63	$1,400-3/95	$2,170-11/97	55%
$20 Liberty	1906-S	MS-63	$1,125-11/95	$1,820-11/97	62%
$20 Liberty	1885-S	MS-64	$9,800-8/95	$14,000-7/97	43%
$20 Liberty	1894-S	MS-64	$4,900-11/95	$6,300-11/97	29%
$20 Liberty	1894	MS-64	$4,600-6/95	$6,300-11/97	37%
$20 Liberty	1895	MS-64	$2,850-6/95	$3,850-11/97	35%
$20 Liberty	1895-S	MS-64	$3,080-6/95	$4,900-11/97	60%
$20 Liberty	1896-S	MS-64	$4,200-2/95	$6,300-7/97	50%
$20 Liberty	1898-S	MS-64	$1,875-7/95	$2,310-7/97	23%
$20 Liberty	1899	MS-64	$1,750-8/95	$2,380-7/97	36%
$20 Liberty	1900-S	MS-64	$4,900-11/95	$5,880-11/97	20%
$20 Liberty	1901-S	MS-64	$5,600-11/95	$7,000-7/97	25%
$20 Liberty	1906-S	MS-64	$2,975-3/95	$5,600-4/97	88%
$20 Liberty	1907	MS-64	$1,750-1/95	$2,660-7/97	52%

MS-63 Average 35%

MS-64 Average 41.5%

Data compiled by Dr. Fred Goldstein.

For example, the total population of all known rare-date U.S. $20 gold coins is approximately 125,000. In 1998 alone 150,000 gold bullion coins were minted globally - in just one month. (That's 1,800,000 for the year!)

So if anyone tells you that a rare coin is not a collectible, when only 5, 10 or 20 pieces remain in existence, they are mistaken. Or if anyone tells you that a bullion coin that is still in production is "rare," tell them you know better. Sadly, unscrupulous coin dealers spread misinformation to confuse the buyer, disgracing the coin profession.

Real Wealth... To Go

Let's talk sheer practicality for a minute. Have you ever seen $100,000-worth or even $10,000-worth of gold or silver bullion coins?

You would need an armored car to transport the bullion holdings of some of my clients. When we are talking about preserving a substantial amount of wealth, bullion coins can be very bulky and difficult to transport without attracting a great deal of attention.

U.S. rare coins on the other hand, are very portable. Owners of rare coins can carry a substantial portion of their portfolio comfortably in a purse or briefcase without raising an eyebrow from onlookers.

Most of the wealthiest investors, who have done their research on which type of gold to buy, have chosen investment-grade U.S. rare coins because they offer the best of everything - liquidity, privacy, profitability and finally, portability. True, they cost a bit more, but they are well worth it over the long term.

2.2 Coin Valuation Factors

U.S. rare coins by their very nature are becoming harder to find every day. The **1903 Barber $ .25 Proof** number of coins originally issued by the U.S. Mint is not always an accurate indication of scarcity.

It is the interplay between the condition, the supply and the demand of a coin that determines its scarcity and market value.

Condition

Perhaps the most important factor in rare coin valuation is the condition of the coin. The grading of coins is performed under exacting specifications that requires specialized training.

Numismatic experts use the *Sheldon Grading Theory,* a numerical scale from Mint State 1-70 to describe the condition. Coin grading takes into account both the obverse (front) and the reverse (back) of each coin to establish an official coin issued by a third-party coin grading service (PCGS, NGC, etc.) There are three major categories of condition that are important: **circulated, uncirculated** and **proof** coins, which are always uncirculated.

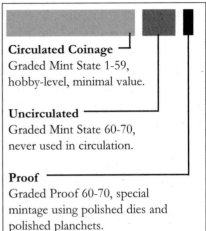

Circulated Coinage
Graded Mint State 1-59, hobby-level, minimal value.

Uncirculated
Graded Mint State 60-70, never used in circulation.

Proof
Graded Proof 60-70, special mintage using polished dies and polished planchets.

Coins graded Mint State 1-59 were circulated and exhibit some wear. Coins graded Mint State 60 and higher are referred to as uncirculated, denoting they exhibit no wear due to circulation among the general public.

Proof coins are a special collector edition with very low mintage that were never designed for circulation. They are minted from highly polished mint dies and planchets (slugs) that are often struck several times to create a crisp, clear image. These specimens are graded on a scale of Proof 60-70.

After the grade is assigned, coins are then placed in a hermetically sealed, tamper-proof plastic holder with the grade and certification permanently displayed by a third-party grading service. The two most popular are Professional Coin Grading Service (PCGS) and Numismatic Guaranty Corporation (NGC).

PCGS and NGC have helped offer both rare coin collectors and investors to acquire rarities with 100 percent confidence of authenticity and the accurate grading of a rare coin. The professionals of these firms carefully examine each coin under the most ideal conditions.

Supply & Demand

The basic free market axiom, known as *the law of supply and demand,*
is a key factor behind the increasing value of high quality collectible
coins over time. As more people enter the coin market, shrinking
product availability causes prices to rise - a sharp contrast to the
decreasing value of the dollar. This benefits U.S. rare coin investors
and collectors by rewarding a long-term strategy, putting time on their side.

Nowhere can the law of supply and demand be seen as clearly than
in the rarest of the rare coins known as "key date" rare coins. The
key date market segment includes issues that have never been available in circulation and rare Proof-only issues such as the 1877 Three
Cent nickel, 1877 Shield nickel, and 1895 Morgan dollar.

Key date coins have been included in many famous collections
because they are almost impossible to find. The dwindling supply
of key date coins means any increase in demand allows the seller to
virtually name their price.

No Competition

When it comes to keeping pace with the rising cost of living
through the last century, neither the U.S. dollar, gold or silver bullion
can compete with investment-grade U.S. rare coins because the U.S.
dollar retains a measly three percent of its buying power. You need at
least 30 times more buying power in your "buck" just to stay even.

For example, let's compare rare coins to bullion during the last 200
years. In 1801 you could buy an ounce of silver for $1 and a half-
ounce of gold for $10. Today an ounce of silver bullion is worth
about $5 and a half-ounce of gold bullion is about $130. So, 200
years later, silver bullion has only five times more buying power and
gold bullion only 13 times more buying power.

In contrast, a 200-year-old silver dollar (1801) is valued at $575
today. A 1801 $10 gold coin today is worth $5,275 in any condition.

That is an average of 600 times more buying power!

Whether you compare bullion and rare coins before the War of 1812, during the good times in the roaring '20s, through the bad times of the Great Depression or right up to this day - collectible rare coins have always outperformed bullion coins.

Does anyone seriously expect bullion coins to miraculously outperform collectible coins in the 21st century? Only a self-serving bullion dealer or their misinformed clients would argue this point.

Times are changing, but collectible U.S. gold and silver coins have passed the test of time. Gold bullion remains one of the most enduring, useful and beautiful precious metals on earth, but when it comes to growth potential, it can't compete with investment-grade U.S. rare gold and silver coins over the long-term.

A $68,000 Difference

Here is an example of a client (Mr. C) who listened to our advice in 1995 and bought high-quality investment gold coins instead of bullion coins. Mr. C invested $107,085 in U.S. rare coins in 1995 and 1996. In May 1999, Mr. C. decided to cash in his portfolio, valued at $146,094, for a net gain of $39,009.

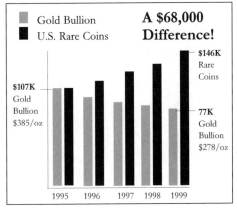

At the time of Mr. C's initial investment, gold bullion was $385 per ounce. At that price he could have purchased 278 ounces of gold bullion ($107,085 divided by $385). By May of 1999, the price of gold bullion had dropped to $278 per ounce. If Mr. C had invested his $107,085 in gold bullion rather than rare coins back in 1995, his portfolio would only have been valued at $77,284 in May 1999 - creating a net loss of $29,801 instead of a net gain of $39,009.

Conclusion

Mr. C's experience clearly supports my premise that all gold is not created equal. Holding the right type of gold coins helped Mr. C turn a potential loss of $29,801 into a gain of $39,009 - that's a $68,800 difference.

Identifying opportunity is a lot easier in hindsight. All gold isn't created equal, but we all have an equal chance to identify new opportunities in today's marketplace.

Back in 1999 many stock investors were bemoaning the lost opportunity when it comes to the raging dot-com stock market. Here's your chance to get in on the ground floor of the next gold rush. I believe that today's U.S. rare coin market has the same type of growth opportunity as the stock market, with far less risk.

Remember, always buy quality because all gold is not created equal.

Learning the basics of buying U.S. rare coins is the next step in our journey to rediscover gold in the 21st century.

3

Introduction to U.S. Rare Coins

Augustus St. Gaudens
designer of $20 Gold piece

Since the early 1950s there has been a growing number of U.S. rare coin collectors. These collectors provide strong demand for high quality U.S. rare coins that are graded gem quality. Passionate collectors thrive on making historical connections to the past and enjoying the artistic beauty of U.S. rare coins. Profit potential is rarely a major factor for the collector.

ABC Dateline recently did a story discussing the rise of U.S. interest in collecting coins and estimated that 40 percent of Americans are now involved in collecting the new Commemorative State Quarters.

Thirty years ago rare coin interest got another big boost based on the growing inflationary concerns and the reopening of free market gold trading in 1971, which began to attract a new breed of U.S. rare coin buyer: the investor.

Initially a relatively small number of rare coin buyers began purchasing coins for inflation protection as well as profit potential. Many have been rewarded well through the years. Today an estimated 25-30 million investors and collectors are active in the U.S. alone.

Although gem-quality coins have had an excellent performance record since 1970, fewer than 10 percent of Americans actually participate in this market. The reasons vary, but most commonly it is the lack of knowledge about coin pricing, grading, dealers, liquidity and volatility, which has frightened away potential buyers until recently.

3.1 Rare Coin Collecting and Investing
Certification, Grading & Population

The introduction of third-party certified grading in the mid-1980s was a major breakthrough in the industry because it enabled retail and wholesale buyers and sellers to purchase U.S. rare coins of a guaranteed condition (grade) that is accepted worldwide.

During the 1990s, standardized rare coin grading and authentication has given the needed confidence to millions of novice coin buyers and investors to enter the market. As mentioned, the two major independent certification firms that have emerged as the leaders are NGC and PCGS.

Dealers, collectors and individuals may submit their uncertified rare coins ("raw") to these two independent organizations for certification. The professionals of these organizations carefully and anonymously examine the coins under ideal conditions.

Each coin is assigned a grade and encapsulated only after passing expert inspection. The grade assigned is backed by an impressive insurance policy underwritten by Lloyds of London.

Mint State grades are awarded to uncirculated coins exhibiting no wear. These grades begin at Mint State 60 and go through Mint State 70, which is perfect. Coins graded lower than 60 are not generally considered investment-grade unless they have a very important date significance.

The four basic criteria for establishing the coin's certified grade are:

1. **Mint strike**
2. **Original luster**
3. **Bag marks**
4. **Eye appeal**

1. Mint strike refers to the sharpness and accuracy of the strike at the mint; a slightly skewed angle of strike will change the appearance of the coin.

2. The original luster refers to the color and toning of the coin and originality of surface. Cleaning of rare coins is taboo because it disturbs the original mint condition.

3. Bag marks on the coins are slight scratches, marks and imperfections resulting from coins pressing against each other in the original mint bags during transportation to banks.

4. Eye appeal is the overall end result of how the coin pleases the eye. So the eye appeal is the overall beauty of the coin, based on the strike, luster and bag marks. Within a specific type and date of a coin, the higher the Mint State rating, the more rare and valuable it is.

The vast majority of coins that are traded in the marketplace today by dealers, collectors and investors are PCGS or NGC certified because these services have eliminated the risks associated with grading subjectivity in the past.

These two grading services keep monthly records on each coin certified. The coin population of any issue is then published. Two respected publications in the industry are the Coin Dealer Newsletter (CDN) and Certified Coin Dealer Newsletter (CCDN).

The lower the population of a particular coin, the scarcer it is compared to other coins of the same type and grade. The population of a coin or the population of its entire type is a helpful statistic in determining a coin's rarity, market value and growth potential.

Where Do Fresh Graded Coins Come From?

After the U.S. government confiscated gold from the public in 1933, almost all the surviving uncirculated $20 Double Eagles were shipped to Europe.

Today the majority of new or fresh grade Mint-State $20 Double Eagle gold pieces still come into the U.S. market from European banks. This steady stream from Europe has been fairly consistent during the last 25 to 30 years, while U.S. demand has had a greater effect on prices.

Collector demand has been steady, although increasing lately, while investor and dealer demand has significantly increased due to inflationary concerns and a bottoming of gold prices. Any disruption of the supply of fresh $20 gold pieces from this primary source has a significant impact on dealer inventories. Therefore, a restricted supply causes dealers to raise prices to be able to meet the demand. Like other markets, supply and demand has greatest impact on prices.

New supplies of certain rare coins, such as Early American coinage or Proof Type coins, generally come from auctions and collections. These supplies are much smaller compared to Double Eagles, meaning demand has a more dramatic effect on prices.

Fundamental economic factors such as oil prices, geopolitical instability, interest rates and disposable income can indirectly affect supply and demand. Therefore, the pricing of rare coins will be determined by a combination of market factors.

Today the rare coin market is growing rapidly and supply is not keeping pace with demand for most rare issues.

Investors and collectors should expect to pay a commission when purchasing U.S. rare coins from a dealer or broker. Certain dealers also charge a commission when repurchasing coins. Always ask your broker about commissions, spreads and the buy-back policy.

This factor is extremely important because commissions can fluctuate between 5 percent and 30 percent, depending on market conditions and the rarity of the coins. Your broker should thoroughly disclose pricing to you, whether the coin is a common-date Double Eagle or an extremely rare, six-figure U.S. Gold Commemorative.

National and Local Dealers

There are more than 5,000 rare coin dealers in the U.S. alone. A smaller local dealer can be helpful for purchasing lower quality (below MS-60, hobby) coins, but national dealers are more experienced handling investment-grade rare coins (MS-63 through MS-70).

Local coin dealers offer an excellent resource for the hobby coin collector. They cater to novice coin collectors interested in circulated coins, but rarely offer high-performance, investment-grade coins. Local dealers can also be a good resource for those interested in small bullion purchases or liquidations for the convenience factor.

National dealers have more capital for larger acquisitions and a larger, more diverse coin inventory. Larger dealers also tend to offer more educational material to help prospective buyers as well as contacts in the broader marketplace to buy and sell coins.

National dealers also have a greater ability to research current market trends and publish them for potential buyers. In searching for a reputable dealer I suggest asking lots of questions. Try to get as comfortable as possible while getting a good education.

Just as a stockbroker may recommend a specific stock due to the company's growth potential and earnings, a coin dealer may suggest a specific coin due to its rarity, pricing or growth potential. The time to get answers is before you purchase stocks or coins, not afterward.

Intrinsic and Extrinsic Value

Many people I talk to have a common misconception that higher rare coin prices must coincide with escalating gold and silver bullion prices. This is not true.

The rare coin market might as well be a distant cousin to gold and silver bullion coins in a down or flat gold market because they respond independently to the day-to-day ripples in the economy.

For example, during the last decade, gold bullion prices have dropped from more than $400/oz. to less than $280/oz. Yet during the same period the prices of many rare-dated $20 Gold Liberties in MS-63 has increased. Each of these coins contains .9675 troy ounces of gold. While the intrinsic bullion value of the coin decreased, the extrinsic or collector value increased. Therefore, the overall value of the coin increased markedly.

During the same time period, lower quality or "common-date generic" coin issues graded MS-60, MS-61 and MS-62 decreased in price as gold bullion prices dropped. Generally speaking, lower quality generic coins with very high populations do not rise in price while gold bullion prices are dropping. However, they do find a bottom rather quickly, as seen when gold dropped below $300 and generic gold coins did not follow.

During the last major bull or growth phase in rare coins, from 1987 to 1989, bullion prices dropped while rare coin prices were rising.

However, in the event gold bullion prices should start to rise, all gold coins should see price increases. Escalating bullion prices can have a dramatic effect on rare coin prices as seen during the 1976-1980 period.

Rare Gold vs. the S&P 500

Let's now compare rare coins to common stock. Chart #2 compares the performance of the S&P 500 Index to a gold coin index from 1970 to 1997.

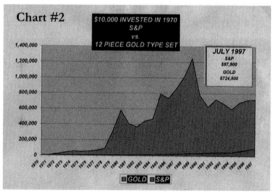

Chart #2

$10,000 INVESTED IN 1970
S&P
vs.
12 PIECE GOLD TYPE SET

JULY 1997
S&P
$97,900
GOLD
$724,500

All data was obtained directly from the CDN at 12-month intervals. The gold coin index consists of the 12 major gold coins (in MS-65 condition) listed in the January 7, 1972 issue of the Coin Dealer Newsletter. In the interest of standardization, the same group of coins was used throughout the course of this study.

In the interest of fairness, the coins utilized are available today to a broad cross-section of collectors and investors. Coins of greater rarity would have turned in a profoundly better performance, but would not necessarily be representative of the norm.

From 1970 to 1997 $10,000 invested in the S&P Index grew to $97,900, almost a tenfold increase. During the same period $10,000 invested in U.S. rare gold coins would have grown to $724,500 - a 72-fold increase!

Chart #2 supports my premise that over the long term U.S. rare coins can be more profitable than the hallowed U.S. stock market.

Many savvy investors and collectors have been utilizing this information for more than 25 years. Buying quality rare coins at the right time and holding them long-term have enabled many to achieve success in this market.

But what about the future? Will the trend toward tangibles grow?

3.2 A Golden Opportunity

Extreme instability is the best description of today's financial markets. Recent headlines reveal that the markets are in for a period of adjustment in the months and years ahead:

"The dot-com bubble has burst." -*USA Today*

"Margin loans to buy stocks have risen to a staggering $243 billion!" -*USA Today*

"U.S. bond market plummets, led by the weakest 30-year bond prices in 17 years." -*MSNBC*

"The price of crude oil topped $30 per barrel... inventories near lowest levels in two decades. Analysts admit leads to inflation." -*WSJ*

From my perspective Wall Street is getting rich while Main Street is struggling to pay the bills. Americans work harder, but earn barely more than in 1980. What is really going on here?

New-economy Internet stocks are like a huge game show that anyone can play, yet history proves that most will lose over the long term. The reason is that they are built on a slim foundation - 50 percent of the NASDAQ capitalization comes from just seven stocks.

Contrarian economist Doug Casey warned in early 2000, "Net stocks are extremely volatile and about to enter a bear market that will deflate prices of the few that survive by 95 percent."

My advice during 1999-2000 was to watch out for the dot-com undertow because it did not have a proven track record or sustainable profits or earnings.

But when would these imbalances be corrected, given that stock speculation is driven by tips, hunches and rumors rather than fundamentals? The answer came slowly starting in March of 2000 when the bull market began to slowly die.

In-and-out trading by unsophisticated investors is gambling, not investing. According to Webster, *invest* means "to clothe, to surround," whereas *gamble* means "to risk, to squander."

What can we learn from history?

"Just look at recent market bubbles for comparison. Gold rose 1,500 percent from '72 to '80. Then it fell back 63 percent. Japanese stocks rose 540 percent from '82 to '90. They too subsequently fell by 62 percent. The NASDAQ has gained 1,200 percent from '92 to 2000. If it falls 60 percent, the index will bottom out in the 1,800 range. That will represent a loss of capital equal to approximately $3 trillion." -Bill Bonner, *The Daily Reckoning,* 2/00

Even *Business Week*'s February 14, 2000 Special Report, "The Boom" strongly cautioned investors: "The strength of the tech sector is covering up growing vulnerabilities, including volatile consumer incomes, high debt levels, and increased dependence on foreign money. All are ingredients for a stock-market led downturn."

This is a major warning to begin to reallocate, or diversify your assets for safety from growing volatility. The next logical question is, "Where can I diversify assets for safety?"

May I suggest a haven of rest for weary stock traders that offers investors safety, privacy and profit potential all in one: investment-grade U.S. rare coins.

The U.S. rare coin market is still the best-kept secret on Wall Street and offers a perfect counterbalance to high-flying stock market speculation.

In the December 1999 issue of *Real Money Perspectives* newsletter, entitled, "Gold: The New Bull Market," announced that "We're entering a bull market, the likes of which we haven't seen since the great glory days of gold in the 1970s, when gold rose over 10-fold."

Since then, more and more voices have begun to chime in confirming the upside potential of gold coins.

For example, CBS *MarketWatch* (2/15/00) announced, "Gold is poised for greatness. Gold has been a stick-in-the-mud for more than a decade. That's all about to change."

The New Bull Market

GOLD!

Recent quotes on Gold by respected newsletter writers:

"I think **a new bull market in gold has started**, not just another cyclical upturn."
-**Douglas Casey,** *Int'l Spectator,* Vol. XX, No. 9

"Technically speaking, **gold is in a new bull market of epic proportions** because of two reasons: First, the spectacular power with which it flew off the bottom, and second, because it has already taken out three of its previous highs dating back to April 1998."
-**Dennis Wheeler,** *Gold Stock Report,* Nov. '99

"The 19 year bear market in gold is over. Finished. Kaput. We're entering **a bull market, the likes of which we haven't seen since the great glory days of gold in the 1970s,** when gold rose over 10 fold."
-**Larry Edelson,** *Year 2000 Alert,* Oct. 20, 1999

"I believe **we are at the beginning of a new bull market in gold,** ... the start of a gold bull market begins slowly and does not roar until a few years down the road. ... be patient."
-**Mark Skousen,** *Forecast & Strategies* Nov. '99

In the 21st century we had better get used to change - not just toward speculation, but back toward fundamentals.

Harry Bingham's *Gold Market Commentary* also foresaw strong fundamentals for gold and explains why:

> **"The equilibrium price of gold today is $600. Primary market conditions are reversing. More and more newly mined gold will be delivered into hedge funds and returned to central bank vaults."**

The best news is that you can take advantage of instability in the gold markets today. Although gold bullion prices could move upward at any moment, based on the growing demand in the physical market, U.S. rare gold coins offer a golden opportunity, even if gold bullion prices stagnate or drop further.

Next we will examine the issue of market timing. Is there a good, better or best time to get started in the rare coin market? The answer is yes, but to understand timing you must consider the market cycles, which is the subject of the next chapter, *U.S. Rare Coin Cycles.*

4

U.S. Rare

Coin Cycles

4.1 The New Bull Market in Gold
The 10-Year Cycle

As a kid in the early 1960s, I used to go to the bank to buy rolls of pennies, dimes, quarters and half-dollars. Little did I know that collecting coins would later become my vocation.

Yes, times were simpler back when Mercury dimes and Indian head pennies could be found in loose change and slipped into my Blue Book coin collection.

Life also seemed more stable because the pace of change was slower and more predictable. Even our money was more stable because you could still go into any bank and exchange a paper promise (a dollar) for the real thing - a silver dollar or two silver half-dollars.

But that all changed in 1965 when even the promise of payment in silver was removed by the Fed. Starting in 1965 our coinage began a gradual descent from 90 percent pure silver to zero today.

I do not think it's a coincidence that ever since the late 1960s the rare coin market has continued to steadily grow in a cyclical fashion. Throughout my 35-year experience in the coin market I've observed two important facts:

1. They don't make historic coins anymore.
2. U.S. rare coins follow a ten-year cycle.

About 20 years ago (1979 -1980) during the Carter administration we had double-digit inflation, double-digit interest rates and gold bullion increases of 721 percent, while numismatic coins escalated 1,222 percent.

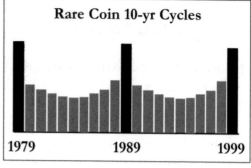

Then suddenly bullion prices dropped with virtually no opportunity to sell out for most investors. Rare coins also dropped, but at a slower rate that allowed enough opportunity for profit taking.

Ten years later (1988-1989) Wall Street first discovered U.S. rare coins and again we experienced a bull market cycle, this time pushing prices up as much as 600 percent. Only those who understood the 10-year cycle were prepared for it.

By the end of 1990 the market began a stair-step decline until 1995. This drop appeared to be an over-correction, as some coins dropped as much as 80 percent, until they hit 14-year lows.

In 1996 the rare coin market stabilized and in 1997 and 1998 brought more than 20 percent average growth. In 1999 many segments of the rare coin market increased by 30 to 40 percent, such as $20 dated Liberty gold coins.

Then came the year 2000 with all of its fanfare. As mentioned, the apocalyptic predictions prompted many gold buyers to opt for bullion gold. Yet it was the rare coin buyers who were the real beneficiaries - despite the falling gold bullion price.

Some U.S. Gold Commemorative coin prices escalated 60 to 75 percent before the end of 2000. Why? Because we are right on target for another 10-year cyclical boom in rare coins.

Only this time, it not only represents a decade, but also a century and millennium economic super-cycle.

Cyclical Volatility

In each of the past ten-year coin cycles, prices have moved up sharply in a relatively short period of time. But the price action of the last few years could be best described as a quiet or accumulation phase of the cycle.

Certain issues, such as U.S. Gold Commemoratives, have been quietly growing at the steady pace of one percent per week for over two years, while others have seen little change or been slightly positive or negative.

In the rare coin market, a downtrend in prices for more than a year is considered a bear market, while price gains of more than 200 percent in two years are considered a speculative or bull market.

This accumulation phase of the cycle is a great time for new coin buyers to enter the market. Prices are generally stable, allowing both investors and collectors to have greater availability of quality coins.

The present accumulation phase has been building since 1997 and is now giving way to a new growth phase where there will be increasingly more publicity to the U.S. rare coin market.

For example, in summer 2000 I was a guest on CNNfn *Business Unusual* and PAX *MoneyWatchTV* to discuss rare coins. I expect this book will also help the process of the new growth phase, and as I (and others) help explain this market to the public in the future.

Everywhere I look I see more and more Americans reaching back into history in their search for meaning and purpose in life. U.S. rare coins are a very important part of our unique American heritage.

It is during this new growth phase that millions of new consumers and investors will become interested in rare coins. We've already

seen a massive year 2000 government and business promotional
campaign as the new Commemorative State
quarters and Sacagawea golden dollar ads
flashed across our TV screens for many
months. I see this as confirmation of the
beginning of a new renaissance for the
U.S. coin market.

As the momentum grows I expect prices
to move sharply higher, leading to the final
"speculative" phase. In this phase coin prices can
rise dramatically in a very short period, similar to the dot-com
stocks in the late '90s, only with rare coins the value is always
tangible - not virtual. During this speculative stage you may decide
to liquidate part of your rare coin portfolio for profit, while main-
taining a core position.

A Millennium Super-Cycle

I feel like a Renaissance man as I look at the extreme imbalances in
our modern geopolitical, economic system which are both on the
verge of historic changes. In my hand is the most valuable asset
mankind has ever known - rare gold.

What will this millennium super-cycle mean to your financial portfolio? One of two things - a giant harvest or a giant headache.

The so-called new economy is now coming back to earth. The
NASDAQ nation is now ready to "get real." The massive ideological
pendulum has begun a historic swing in the direction of morality,
freedom, liberty and justice - all of which draw strength from real
assets like gold, land, livestock and all other commodities.

Consider these key economic factors that support my premise:

**1) Gold bullion is highly undervalued, yet some rare coins are
up more than 100 percent since 1999.**

2) Stock Indexes globally are growing more volatile, fueled by wild stock market speculation and falling productivity.

3) Global economic instability is rising, leading to a major debt-liquidating crescendo.

The cumulative millennium super-cycle effect is unpredictable, perhaps even dangerous. However, in every approaching crisis is also a hidden opportunity. I fully expect rare coins to exceed the performance of the last two cycles.

The only logical reasons for not buying U.S. rare coins at this time are either no liquidity or lack of knowledge. Lack of available funds cannot be helped, and I never encourage going into debt regardless of how appealing a market appears.

The lack of knowledge however, is no longer an acceptable excuse in my book. We learn what we need to get what we want in life.

Sharing knowledge and wisdom is the crux of my business. Education is freely delivered by my brokers over the phone, on our web sites, on the airwaves and in print, because I'm committed to help Americans get educated about how to reach their financial goals. What about rare coin downside risks?

A Flower in a Forest

I view the U.S. rare coin market as a flower in a forest. It is small yet beautiful, and only visible by looking beyond the trees. It is often overshadowed by the looming stock market. However, once it is discovered the investor appreciates its value.

The rare coin market is very thin compared to Wall Street. It is estimated that the total capitalization of the U.S. rare coin market is approximately $20-30 billion - small potatoes compared to U.S. stocks and bonds capitalized greater than $25 trillion.

If $100 million were invested into the Dow Jones Industrials in one day, it would have little effect on prices compared to the same amount invested into the U.S. rare coin market. Due to significantly smaller numbers and the relative value of coins compared to stocks, this same dollar investment would have a dramatic effect on coin prices. The volatile and fragile nature of the rare coin market must be taken into consideration by serious buyers.

Astute stock investors have had success buying larger, well-capitalized blue chips like IBM and GE. However, successful coin buyers (i.e. Pittman and Eliasberg) have made huge profits buying the rarer, more thinly capitalized issues.

Today's rare coin buyer has much to learn from expert collectors of the past about how to utilize the beneficial aspects of volatility to maximize gain and minimize loss. As we begin to view time as our friend, we will discover that rare coins are one of the best long-term capital preservers today.

Of course past performance is certainly no guarantee of future performance, and my analysis of rare coin market cycles is by no means scientific. For that reason, working with a knowledgeable broker enables both the rare coin collector and investor to better understand ever-changing market conditions.

Coin Liquidity & Commissions

U.S. rare coins are very liquid worldwide. Most coins are easily sold for cash at any of the 5,000 dealers in the country. Certain expensive coins may be less liquid than others and selling them at auction may be the optimum means of liquidation.

Swiss America, as well as other large national companies, facilitates clients who need to sell coins through in-house liquidation or at auction. Reputable dealers will send their clients' coins to the auctioneer for photography and marketing prior to the auction.

The larger national companies generally provide a two-way market by offering to buy back, or repurchase, coins sold to their clients. Commissions charged at liquidation are at the discretion of each dealer.

For example, my firm does not charge the customer a commission upon liquidation. Broker commissions on the sale of U.S. rare coins will vary from as little as 5 percent to as much as 30 percent, depending on the firm and the product.

Everyone wants to get the lowest price on the best product, but buyers must learn enough about the product before purchasing or the odds are that you might be disappointed.

I recommend that U.S. rare coin investors plan to hold their coins for a minimum of three to five years.

In certain cases, a longer or shorter holding period may be necessary. Serious collectors are notorious for buying rare coins with no intent ever to sell them.

This growing collector market thinks inter-generationally, and most often will pass on coin collections to their heirs. I think we can all take a lesson from these collectors in light of the speculative and short-term thinking patterns of the 1990s.

4.2 Gold: The Price is Right

In all the years that I've been in the business of buying and selling U.S. rare coins, I've yet to have a month go by that a customer hasn't asked, "Was the price I paid the *right* price?" This is not a simple yes-or-no question.

How is a coin's real value determined? What publications or services provide accurate prices? Did I pay the best price for my coins?

These are some the most commonly asked questions by prospective coin buyers. But where can a coin buyer get reliable answers? I recommend The *Coin Dealer Newsletter* (CDN) which is the most popular guide to pricing. *CDN* has published a weekly report on the coin market for 35 years and is the accepted standard by dealers.

The *Coin Dealer Newsletter* is often referred to as the "greysheet," which serves both dealers as well as the general public. Along with the weekly report, *CDN* also provides quarterly and monthly supplements that detail each individual coin.

The greysheet is set up in a Bid/Ask format and reflects national dealer-to-dealer wholesale prices on a sight-seen basis (lowest price reported).

The greysheet, although a great help in establishing value, is not the final word. Many coins trade for well above greysheet prices - in some cases, even multiples of the sheet - and some trade for well below the sheet. The greysheet gives us a general feel for week-to-week market movement.

In early 1986, when independently "certified" coins started to trade actively, *CDN* launched the *Certified Coin Dealer Newsletter* (CCDN) which began reporting the price activity of certified coins. This report quickly became known as the "bluesheet." This report reflects national dealer-to-dealer wholesale prices on certified coins that trade on a sight-seen basis.

The Official Red Book - A Guide to United States Coins has been published yearly for more than 50 years. The Red Book is a very useful educational tool for the new hobbist and rare coin collectors and investors to learn about a coin's history, original mintage and estimated market value. Because the investment-grade coin market moves on a weekly basis, the greysheet is the preferred price indicator used in the U.S. rare coin marketplace.

Another price factor is tracking the coin auction prices realized each week. A coin dealer must negotiate with wholesalers almost daily. After all this... the true market value of a coin can be established.

But coin buyers don't need all of these services and reports to determine the value of their coins if broker trust is established. If a buyer wishes to track a coin's progress independently, he may want more than just another dealer's opinion.

Sight-Seen & Unseen Coins

In the coin market, price does not automatically equate to value - such as a car purchase or buying a six-pack of Coke. Not all coins are created equal. So-called "sight-seen" coins are priced differently than "sight-unseen" coins. In other words, a coin that is ugly won't sell for the same price as a pretty coin.

For example, David Hall, the founder of PCGS, will tell you that two coins - both graded MS-65 - can have very different values. Yet, I've received calls from clients who have been told they can buy the "same" coin they bought from my firm from another coin dealer for far less than what they paid us. When I inquire exactly who is offering them the "discounted" coin, their voice becomes silent. It seems top secret.

Experience has taught me that once I speak to the other dealer involved (if we are allowed) I discover that they thought the client said Saint-Gauden, not Liberty. Or they were talking about their generic price when the client has a Dated coin, which is always more valuable. My personal favorite is, "We could sell you that coin for $200 less, but we don't have any right now. Perhaps you'd be interested in our special of the week?"

I have traded with more than 35,000 clients in the last 20 years in the coin business. Here's the bottom line: you get what you pay for. There is no free lunch out there. Don't get me wrong, you may find a similar coin, but it won't be the exact same coin.

As you explore further, you'll also find out that the service of the company behind the coin will vary greatly. For example:

• Does the company have a buy-back policy behind it?

• Have they been through the ups and downs of the markets?

• Was it a sight-seen coin that was inspected by three knowledgeable numismatists, or was it looked at by a person who just started in the coin business?

• Was it one coin of a group of five in which four others were rejected due to copper spots or other detractions that could keep the coin from having optimal market appeal to future buyers?

• Are you truly comparing apples to apples?

I repeat, no two rare coins are created equal. A rare coin is not a commodity like sugar, wheat or bullion gold. Have you ever seen a 100-ounce bar of gold? Some are ugly and beat up, some are brand new and shiny, but they both trade for the same value.

This is not true in the numismatic coin market. The way a coin looks - *the eye appeal* - as well as the way it is graded is very important when making your comparisons.

Frankly, unless the other dealer has the coin you bought right in front of him, he can't make the comparison correctly. He simply can't see the coin through the telephone. Even if you went to his shop, he probably doesn't have another coin to compare it with that is the exact grade or date.

Often, coin shops will try to use another date or grade coin, which then invalidates the comparison. My point is that it's easier for many to believe they made a bad trade than it is for them to believe that they bought a great coin at a fair price. We always seem ready to believe a bad report a lot faster than we do a good report.

Sight-seen simply means that the buyer has had a chance to view the coin before he or she buys it. Comparing a sight-seen priced coin with a sight-unseen coin can represent a big difference in value.

Sight-unseen prices are the prices that dealers will pay for coins they have never seen and have no ability to return if they don't like the quality or look of the coin (generally the lowest price).

Simply put, sight-unseen coins are generally bottom-of-the-barrel coins that most reputable dealers reject - or generic coins that are just used to fill orders, without regard for quality. Many newcomers to the market will buy the lower-priced coins and hold themselves out as the "real market" on the coin. Baloney!

Speak to anyone knowledgeable in coins, and they will tell you to always pay a little more and get the better quality coin.

For years David Hall would recommend only MS-65 or better. Why? Simple. He knew that you should always buy the best you can afford.

Many times when comparisons are made, dealers are quoting a sight-unseen (lowest price) against a sight-seen (highest price) coin - making the client think he didn't receive value. When we make the distinction, all of a sudden the coins are neck-in-neck on pricing.

I have personally seen coins from other dealers that were downright ugly being compared to my firm's coins. In one instance, a coin offered by another dealer had so many copper spots that it sold for 15 percent below the current market price.

Six Steps to Pricing

Here are six very practical steps that can help you to determine "the price is right" when buying coins.

Following these steps could save you money and heartache later.

1. Always buy from a recognized national dealer or broker who has the experience of the ups and downs in the market - a dealer with a minimum of a 10-year track record.

2. Be sure the coins can be bought and delivered at the prices quoted. In other words, don't just buy the quote, buy the coin. I can sell coins 20 percent, 30 percent, even 75 percent less than the other guy if I don't have them. Make sure the dealer can actually deliver the coin at the price quoted.

3. Buy only sight-seen coins with a reasonable inspection period. Make sure you have the option of exchanging the coin if you don't like the way it looks. Also, be sure that the coin you buy has been bought by the dealer on a sight-seen basis and has been inspected for copper spots and other detractions by at least two numismatists.

4. Be sure the dealer offers you a two-way market. A reputable dealer will not only sell you the coin, but will also buy it back. Closely examine the buy-back policy. Make sure the buy-back price does not vary on a basis of quantity. Also inquire on the time factor for repayment. Most reputable companies will settle a trade within 72 hours.

5. Truthful disclosure of risk. Be sure the dealer has presented both the upside and downside risks associated with the coins you are purchasing and discussed the commissions. Short-term versus long-term positioning should be explained and understood.

6. Buy from a dealer that you have complete trust in. Check their references. Most importantly, ask many questions and get satisfying answers. Get facts, not hype.

Golden Discrepancy Trading

In recent years different segments of the numismatic markets have headed in opposite directions. Investment-grade coins continue to escalate, despite the lack of performance of gold, while the lower end of the numismatic gold market remains flat or down.

Many investors bought bullion or lower-end numismatic gold as a hedge against the potential of crisis or severe economic or financial market reversal. I have developed a strategy to help my clients take advantage of this golden discrepancy.

History has shown that the investment-grade rare coins have a more dependable pattern of growth than the generic market (bullion, AU, MS-60 to MS-62 Liberty gold, MS-60 to MS-63 Saint Gaudens), and a shift in holdings would be advantageous to the client.

Many of my clients have shifted their strategy from owning gold strictly for protection, to a more aggressive posture with a focus on both growth and protection.

With the prices of bullion and low-end coins dropping as much as 25 percent, the conversion to higher-end coins that are up 25 percent or more can result in a 50 percent differential - a strong motivator.

More importantly, if previous market highs (1989) are achieved in the next bull market, this disparity can be as great as a 500 percent difference. Unless gold was bought strictly as a hedge and growth is not a factor, there is little reason to hesitate on upgrading your gold portfolio.

Changes in market conditions sometimes warrant a shift in directions when the numbers are significant. Your bullion and low-end gold can be put to work for your benefit (including tax advantages) when the timing is right.

For further explanation about gold ratio trading, call your broker to discuss the current market conditions and your options.

Whether you are motivated to invest in gold coins for protection, privacy or profit potential I recommend starting off small and testing the broker and market. Those who treat U.S. rare coins with due consideration will profit in many ways, including a sound financial return on one's investment of time and money.

In 2001 I expect certain U.S. rare coins to outperform virtually every other non-leveraged market based on the leading indicator - U.S. Gold Commemorative coins. In Chapter 5, I will explain *The Amazing Story of Gold Commemoratives* in detail and I think you'll understand more clearly why I'm so bullish on truly rare coins.

5

The Amazing Story of U.S. Gold Commemoratives

5.1 A Mirror of History

U.S. Commemorative coins are a mirror of history and art that tell the American story throughout the past 200 years, reflecting social, political and economic triumphs and struggles.

In early 1999 I began to closely examine the Gold Commemorative market and discovered that it was extremely undervalued. I immediately issued a market alert to all my clients and then compiled the research into a special report and published it for all to read.

By March of 2001 this small market sector had grown an average of 35 percent for two consecutive years - while most other coin sectors of the U.S. rare coin market remained relatively flat.

I view this recent price movement as just the tip of the iceberg. As of July 2001 most Gold Commemoratives are still only a fraction of their market highs from the last bull market. That means there is still plenty of room for growth over the next few years.

Let's look at these golden treasures in more detail to better understand what is behind this amazing market phenomenon. First some background.

In the U.S., the creation of a commemorative coin requires an Act of Congress to be passed. U.S. Commemorative coins are not created for circulation, they're designed for collecting and minted in recognition of special events. They're sold by the U.S. Mint to collectors at a premium compared to coins minted for circulation.

Artists are commissioned to bring exquisite design to each commemorative coin, creating a new and unique historical icon. Nine types of U.S. Commemorative Gold coins have been issued throughout the last two centuries, five in one-dollar denominations.

Originally there was not much distinction between regular currency and commemorative coinage, with roots as far back as ancient Greek coins that often commemorated many things: sacred animals, local industry, victorious battles or glorified cities and heroes.

From 1934 to 1936, FDR was very interested in commemoratives and almost every single state had their representatives pushing bills for new commemorative coins. At that time there were some odd suggestions, such as a coin representing the Cincinnati Musical Center, which turned out to be entirely fictitious. At the same time, a few very important events were missed, like the end of the Civil War and the signing of The Bill of Rights.

The first souvenir U.S. Gold Commemorative minted, (not for circulation) was the Louisiana Purchase Exposition coin in 1903. Two varieties were issued: one with a portrait of Thomas Jefferson, who was president when the Louisiana Territory was purchased from France; the other depicting President William McKinley.

These original Commemorative coins were designed by artist C.E. Barber, and paved the way for many to follow, each with its own unique story and attributes.

For instance, the Panama-Pacific Exposition quarter eagle minted in 1915 was the first $2.50 Gold Commemorative. For the same exposition, a round and an octagonal $50 gold piece was also issued - the only octagonal coin ever issued by the Mint. The Philadelphia Sesquicentennial (150th anniversary) coin is the only other $2.50 Gold Commemorative. About 46,000 were issued in 1926.

For anyone who considers himself a history buff, Commemoratives offer interesting lore - including scandalous and even sometimes criminal activity connected with the making and distribution of certain coins.

For example, in regard to the 1924 Hugenot, the Federal Council of Churches of Christ used the coins as a fund-raiser - a violation of the first Amendment. This type of a scandal serves more than just a story to tell grand kids, it makes the coins more attractive to collectors, therefore increasing their value. They become memorabilia of events in history that underlie the events for which they were created in the first place.

Timeless Treasures

U.S. Gold Commemoratives have an excellent track record of growth during the last 50 years - averaging more than 1000 percent! See the chart below.

U.S. GOLD COMMEMORATIVE PRICE HISTORY (1950-1985)

Gold Commemorative Coins	1950	1960	1970	1976 (low)	1980 (high)	1985
1903 LA Purchase Jefferson $1	$13.50	$45	$75	$325	$1,900	$1,300
1903 LA Purchase McKinley $1	$13.50	$45	$75	$325	$1,900	$1,300
1904 Lewis and Clark $1	$55	$200	$300	$835	$5,000	$3,000
1905 Lewis and Clark $1	$52.50	$200	$310	$825	$5,000	$3,000
1915S Pan Pac $1	$9	$30	$67.50	$210	$4,000	$1,600
1915S Pan Pac $2 fi	$50	$150	$300	$1,100	$11,000	$5,000
1916 McKinley $1	$12.50	$37.50	$70	$220	$1,825	$1,450
1917 McKinley $1	$15	$40	$135	$290	$2,000	$1,700
1922 Grant With Stars $1	$22.50	$110	$250	$695	$4,150	$1,900
1922 Grant No Stars $1	$27.50	$130	$262.50	$680	$4,150	$1,900
1926 Sesquicentennial $2 fi	$12	$30	$57.50	$175	$1,175	$1,100

Here is a brief description of the U.S. Gold Commemorative eleven coin series provided by Kevin Lipton of KLRC of Beverly Hills.

The 1904 and 1905 Lewis and Clark gold dollars were designed to commemorate the 100th anniversary of their exploration of the northwest United States. These two issues are the rarest of all the $1 Gold Commemoratives with only 10,000 of each minted.

The 1915-S $1 and $2½ Panama Pacific gold coins were struck to commemorate the opening of the Panama Canal, the greatest engineering achievement of its time. The coins were offered with a half-dollar and the rare round and octagonal $50 "slugs" at the Panama Pacific Exposition held in San Francisco in 1915.

One could purchase the individual coins or a complete set in a custom box or copper frame. Only 15,000 one-dollars and fewer than 7,000 of the $2.50 were minted.

In 1916 and 1917, gold dollars were struck to commemorate President McKinley. The obverse contained a portrait of the assassinated President, while the reverse portrayed his memorial building in Niles, Ohio.

1922 marks the year the U.S. minted a coin commemorating the centennial birth of Ulysses S. Grant, the great Union general and post-Civil War president. Just 5,000 of each variety were made; the first with a star on the obverse, the second with no star, which is their main distinguishing characteristic.

In 1926 the final gold commemorative was struck for the sesquicentennial anniversary of our Declaration of Independence. The obverse portrays Lady Liberty, while the Freedom Hall in Philadelphia is pictured on the reverse. This fabulous $2½ gold coin has a mintage of 46,000 and is very difficult to obtain in gem condition.

Supply and Demand

U.S. Gold Commemorative coins have extremely low original mintages - an average of 12,000. Original mintages range from a low of 5,000 (1922 Grant $1) to a high of 45,019 (1922 Sesquicentennial $2.50). The total known population in MS-65 grade averages only 557 specimens per issue and in MS-67 grade only 58 coins per issue.

Because of their low original mintages and low surviving population, U.S. Gold Commemorative coins have broad market appeal to both coin collectors and investors. As a result of the limited supply and growing demand, the price levels began moving up in 1999.

Charts indicate that upward price movement began to accelerate in the second quarter of 1999 and continue even as this book is going to press.

Public Appeal of Commemoratives

At the onset of the new millennium, the public appeared to become more nostalgic. The recent growth of interest in other collectibles such as fine art, antiques, stamps and classic automobiles clearly supports this trend. For example, in recent years we have seen the rise of cable television auction-formatted programs featuring American collectibles. America's appreciation and interest in history is rising.

In 2000, the U.S. government's nationwide promotional campaign for modern U.S. Commemorative coins gave these gems an unexpected public boost. The release of the Year 2000 Sacagawea Golden Dollar and the new U.S. State Commemorative Quarters mark the first time in many years that coin collecting has been elevated to a recommended national pastime.

Especially helpful to the U.S. Gold Commemorative coin market was the multi-million-dollar mass media campaign promoting collection of all 50 State Commemorative Quarters on a colorful U.S. map. This has increased the interest and demand for all types of U.S. coins, but has advanced the commemorative segment the most.

Because the market for U.S. commemoratives has been on an uptrend during the last eight consecutive quarters, it is now considered a bull market.

The consistency of growth during the last two years indicates that increasing numbers of collectors and investors are assembling complete U.S. Gold Commemorative sets with all 11 specimens.

This trend toward conversion of rare coin investors to coin collectors who are interested in acquiring complete sets provides an ever-growing foundation for the U.S. Gold Commemorative market in the future.

An Insider's Perspective

Let's now look at why U.S. Gold Commemorative coins have become the best performing investment-grade U.S. rare coins in the marketplace from an insider's perspective.

Both the *Certified Coin Dealer Newsletter* and the *Coin Dealer Newsletter* have consistently documented the amazing price growth of Gold Commemoratives during the last two years.

For example, U.S. Gold Commemoratives have been featured by the *CCDN* and *CDN* in more than 40 issues and were in the headlines 20 times between March of 1999 and April of 2001. Spot a trend?

"Gold Commemoratives Continue to Increase," read the headline of the December 15, 2000 *Coin Dealer Newsletter*. "As we have written for quite some time in 2000, Gold Commemoratives are facing very strong dealer demand. Plus signs are dominating nearly every issue from this relatively short numismatic series. Demand is strongest for coins properly graded Mint State 64 and better."

Déjà vu, a year earlier the *CDN* headline read, "Gold Commemoratives Hot" (December 10, 1999). Here are a few examples of the press coverage that U.S. Gold Commemoratives have had in 1999-2001.

DATE PUB   HEADLINE and QUOTE

6/18/99 CCDN   GOLD COMMEMS IMPRESSIVE
Gold Commemoratives have come alive this week. We see strong bidding for MS-65 and MS-67 condition.

7/2/99   CCDN   GOLD COMMEMS LEAD MARKET
The Gold Commemoratives are leading the way in the Commemorative market. Dealers are in strong need of Gold Commemoratives.

12/10/99 CDN   GOLD COMMEMORATIVES HOT
Mint State Gold Commemoratives are realizing strong support and demand.

1/21/00 CCDN   LEWIS & CLARK COMMEMS ACTIVE
Promotion has targeted the 1904 and 1905 Lewis & Clark Exposition $1 Gold Commemoratives.

4/7/00   CCDN   GOLD COMMEMS COMMAND ATTENTION
Gold Commemoratives are the stellar performers this week... bids are increasing for a whole host of dates and grades.

9/25/00 CCDN   GOLD COMMEMS SOARING
Gold Commemoratives are very impressive this week in the rare coin market. Dealers are actively increasing their "sight-unseen" bids for these special Commemoratives. This is what eager buyers are hoping will occur. Until this happens, price may only have one direction for the immediate future of Gold Commemoratives.

12/15/00 CDN   GOLD COMMEMORATIVES CONTINUE TO INCREASE
As we have written for quite some time in 2000, Gold Commemoratives are facing very strong dealer demand. Plus signs are dominating nearly every issue from this relatively short numismatic series. Demand is strongest for coins properly graded Mint State 64 and better.

1/12/01 CCDN GOLD COMMEMS LEAD THE WAY

As one of the most active areas of the 2000 market, Gold Commemoratives continue to lead the way as we begin 2001. Any classic gold Commem grading MS-64 and above can be easily sold.

2/16/01 CDN GOLD COMMEMS REMAIN STRONG

Classic Gold Commems remain strong, availability of fresh material is virtually nonexistent.

4/06/01 CDN INDIAN & COMMEM GOLD SKYROCKET IN FIRST QUARTER, CENTRAL STATES SHOULD SATISFY SOME DEMAND

Quote: Gold commemoratives are receiving the headlines again this week. The demand is so strong that the current supplies are just too insufficient to keep pace. Consequently, dealers are left with little choice but to increase their sight unseen bids for these Gold Commems. Plus signs are omnipresent in this short series. Demand is targeting grades MS-62 thru MS-66. The chart indicates the sharp increases for certified Gold Commems.

GOLD COMMEM LEADERS	GRADE	INCREASE
1904 Lewis & Clark Expo. $1	NGC MS-62	+ 23.08%
1916 McKinley Memorial $1	PCGS MS-63	+23.08%
1917 McKinley Memorial $1	NGC MS-62	+21.62%
1903 LA Purchase/Jefferson $1	NGC MS-63	+15.56%
1915 S Panama Pacific Expo. $2 fi	NGC MS-63	+14.67%
1903 LA Purchase/McKinley $1	NGC MS-63	+14.29%
1903 LA Purchase/Jefferson $1	NGC MS-64	+11.56%
1903 LA Purchase/Jefferson $1	PCGS MS-64	+11.56%
1917 McKinley Memorial $1	PCGS MS-63	+10.66%
1915 S Panama Pacific Expo. $2 fi	NGC MS-62	+10.43%

Need I say more? The opportunities for successful collecting and investing in investment-grade U.S. Gold Commemoratives are better today than at any time in the last 15 years. Buying rare coins can be a rewarding experience provided the buyer is armed with the right attitude and background knowledge of the market.

In my opinion U.S. Gold Commemorative coins represent an excellent value at today's price levels and should be acquired while market prices are at such favorable levels.

If you've been waiting for just the right opportunity to diversify some of your paper or non-performing gold into quality collectible gold, then wait no longer. Gold Commemoratives offer the safety of gold ownership along with the privacy and profit potential of a rare collectible. It's a win-win situation.

While other areas of the numismatic market continue to track bullion as a non-performing asset, Gold Commemoratives remain an independent market-mover and should continue this trend for years to come.

As mentioned earlier the current market prices for investment-grade Gold Commemoratives are still only 25-35 percent of the previous market highs in 1989. This indicates a strong growth potential in comparison to other segments of the U.S. rare coin market. When an unbiased industry publication like *Coin Dealer Newsletter* consistently speaks so highly of a market - like Gold Commemoratives - it confirms a trend and limits downside risks.

U.S. Gold Commemorative Scorecard (1999-2001)

The following chart shows the stellar performance of the 11 U.S. Gold Commemorative coins between March 31, 1999 and April 23, 2001 in MS-64 through MS-67, which I believe offer the most growth potential.

UNITED STATES GOLD COMMEMORATIVES 1903 - 1926

Coin Description Mintage	Grade Mint-State	Retail 3/31/99	Retail 4/23/01	% Increase 24 Months	1989 Retail
1903 LA Purchase Jefferson $1	64	**$1,201.00**	**$1,900.00**	58%	**$6,545.00**
17,500 Minted	65	$2,115.00	$3,735.00	77%	$11,555.00
	66	$3,665.00	$6,205.00	69%	$21,850.00
	67	$8,800.00	$14,300.00	63%	$66,200.00
1903 LA Purch McKinley $1	64	**$1,140.00**	**$1,835.00**	61%	**$6,199.00**
17,500 Minted	65	$2,225.00	$4,090.00	84%	$12,625.00
	66	$3,525.00	$6,415.00	82%	$24,325.00
	67	$10,725.00	$15,730.00	47%	$69,300.00

UNITED STATES GOLD COMMEMORATIVES 1903 - 1926

Coin Description Mintage	Grade Mint-State	Retail 3/31/99	Retail 4/23/01	% Increase 24 Months	1989 Retail
1904 Lewis and Clark $1	64	**$3,284.00**	**$5,500.00**	67%	**$12,705.00**
10,000 Minted	65	$6,200.00	$12,400.00	100%	$33,880.00
	66	$9,500.00	$19,305.00	103%	$61,900.00
	67	$22,880.00	$34,320.00	50%	$130,900.00
1905 Lewis and Clark $1	64	**$5,236.00**	**$8,035.00**	53%	**$19,635.00**
10,000 Minted	65	$14,300.00	$25,740.00	80%	$61,000.00
	66	$25,740.00	$40,040.00	56%	$112,400.00
	67	Very Rare			$154,000.00
1915-S Pan Pac $1	64	**$963.00**	**$1,340.00**	39%	**$6,045.00**
15,000 Minted	65	$2,115.00	$3,735.00	77%	$11,550.00
	66	$3,665.00	$6,135.00	67%	$41,425.00
	67	$11,800.00	$17,160.00	45%	$109,325.00
1915-S Pan Pac $2½	64	**$3,542.00**	**$5,780.00**	63%	**$11,011.00**
Less than 7,000 Minted	65	$4,800.00	$7,755.00	2%	$17,225.00
	66	$6,345.00	$10,725.00	69%	$40,800.00
	67	$17,875.00	$24,310.00	36%	$97,500.00
1916 McKinley $1	64	**$878.00**	**$1,305.00**	49%	**$5,313.00**
15,000 Minted	65	$2,115.00	$3,735.00	77%	$13,244.00
	66	$3,735.00	$6,700.00	79%	$28,000.00
	67	$12,870.00	$19,305.00	50%	$92,400.00
1917 McKinley $1	64	**$1,478.00**	**$2,750.00**	86%	**$6,892.00**
5,000 Minted	65	$2,610.00	$4,795.00	84%	$13,700.00
	66	$4,795.00	$7,755.00	62%	$38,100.00
	67	$19,300.00	$24,310.00	26%	$84,700.00
1922 Grant $1 (No Stars)	64	**$1,879.00**	**$3,665.00**	95%	**$9,086.00**
5,000 Minted	65	$2,960.00	$4,655.00	57%	$13,275.00
	66	$3,800.00	$6,345.00	67%	$26,750.00
	67	$9,300.00	$13,155.00	41%	$96,250.00
1922 Grant $1 (With Stars)	64	**$2,187.00**	**$3,805.00**	74%	**$9,279.00**
5,000 Minted	65	$2,280.00	$4,655.00	104%	$13,090.00
	66	$3,535.00	$6,345.00	79%	$22,475.00
	67	$7,050.00	$11,440.00	62%	$54,650.00
1926 Sesquicentennial $2½	64	**$980.00**	**$1,270.00**	30%	**$6,237.00**
46,000 Minted	65	$3,800.00	$5,215.00	37%	$36,325.00
	66	$18,590.00	$25,740.00	38%	$92,400.00
	67	Very Rare			

U.S. Gold Commemoratives experienced a very dramatic price rise during the last bull market in 1988-89 as well. The market value of some Gold Commemoratives today are still near 1985 prices, suggesting that this market may have a prolonged period of growth.

5.2 Commemorating a Timeless Strategy

The John Jay Pittman Story

On October 21, 1997, six days before the Wall Street rollercoaster nearly derailed, approximately 200 people quietly gathered in Baltimore, Maryland for the auctioning of one of the most exquisite coin collections in history, yielding an astonishing 30,000 percent return!

Friends and family gathered in awe, watching coin connoisseurs spend up to $465,000 for one gold coin. They shook their heads in disbelief, saying to one another, "He was so right about coins!"

John Jay Pittman was a man as unique as the coins he collected, yet most saw him as just an average guy. He wasn't a multi-millionaire, nor did he inherit his coins from a rich family member.

Pittman was a humble man of modest means, earning between $10,000 and $15,000 per year working as a chemical engineer for Kodak. He did, however, have a great love for his coins and some years invested up to half his salary in rare gold and silver coins.

Mr. Pittman spent years buying the best, the most historically significant coins his budget would allow. It is estimated that Mr. Pittman invested about $100,000 during his lifetime in rare coins.

The first section of his holdings were sold at auction for $11.8 million (WOW! What a return!). But the best came six months later. The second portion of his coin portfolio sold for over $18 million, bringing the total market value of his collection to $30 million! For those without a calculator handy, $30 million divided by $100,000 is a 30,000 percent growth on his coin investments.

I love to read about coin buyers like Mr. Pittman because he's a perfect example of a man who understood the wisdom of buying quality collectibles and holding them long-term - knowing their value would multiply upon resale, in this case 30,000 times.

Now you might be saying to yourself, "Good for Mr. Pittman, but that wouldn't happen to me." Oh no? Remember, Mr. Pittman didn't have large amounts of money to buy coins, he also raised a family. But, instead of buying stocks and bonds (popular wisdom) he preferred to collect real wealth in the form of rare coins that he could enjoy and keep in his own possession. If you follow Mr. Pittman's model you too could see your coins skyrocket over the next 10, 20 or 30 years.

Here is the profile of a satisfied coin investor or collector that I've seen time and time again: people who buy coins for the right reason, at the right time, from the right broker, and then allow time to become their best friend. John Pittman didn't allow outside influences around him to deter him from assembling what he knew would ultimately become a tangible fortune.

Yes, John Jay Pittman desired to own assets that weren't someone else's liability, and his faithfulness to that task was greatly rewarded. He knew deep down that each pristine coin he acquired would withstand the test of time, and he was right.

Do you think rare, high-quality U.S. coins bought today will have any less value in the future? I don't. Especially when we examine the major 2001-02 economic trends, which is the subject of Chapter 6.

6
Global
Economic Trends

6.1 The Next Wave

Being a lover of the ocean, surfing and ocean sports all my life, I've discovered how important it is to "get up to speed" with an oncoming wave or risk having your nose slammed into the sand… or worse.

That which is true about the nature of waves is also true about the nature of economics. America has just gone through an unprecedented period of financial growth fueled by the longest bull market in stocks in recent history. But the tide is turning.

The total losses calculated from the dot-com carnage illustrate that the year 2000 brought a different kind of computer bite than was expected. An estimated $5 trillion (half our total GNP) was jettisoned into cyber-space in the year 2000.

In early 2001 many analysts began to warn of darkening financial clouds and a possible recession. Students of economic history know that bear markets always follow bull markets. Super-bull markets are usually followed by super-bear markets - it's just how the economic cycles work.

Between 1997-2000 I warned investors of a looming stock market crisis because the so-called new economy was driven by such wild speculation.

From March 2000 through March 2001, the NASDAQ lost more than 60 percent of its value. But what if the worst still lies ahead? A not-so-soft landing.

In 2000 many investors began to learn a very expensive lesson: the so-called financial experts are often wrong in calling market tops and bottoms.

My goal is to help you develop an "exit strategy" to move a small portion of your assets from high-volatility stocks into high-performance U.S. gold coins.

Over the years I've noticed that investors equipped with the bigger picture tend to have better discernment on the difference between investing and gambling.

Wall Street: Investing or Gambling?

On radio interviews I'm regularly asked about my view of the stock market. In 2000 I only published one article on Wall Street and one article on the NASDAQ bubble.

I consistently warned investors that the powerful bull market of the last 18 years had lulled them into a false sense of security, and that novice investors are prone to buy stocks based on tips rather than fundamentals like earnings and profits. It is not really investing at all - it's gambling.

I remember a *Sixty Minutes* interview with an online gambling outfit that said they modeled their business after the Wall Street speculation model, claiming "stocks are just legalized gambling."

Rather than fattening the bull for slaughter, my clients diversified their portfolios and some discovered that their rare coins had appreciated as much as 90 percent in 90 weeks. Yes, my clients were prepared for the dot-com information undertow and found a buried treasure in the form of U.S. Gold Commemorative coins.

Investors who took heed and diversified a portion of their assets into collectible U.S. rare coins discovered that they performed exactly as they should have, serving as the perfect counterbalance to the wobbling stock market.

In 2001 and beyond, as American investors rediscover the value of historic and rare coin, I believe that this product will grow exponentially. Why? The single principle that has been my central message for two decades is simple but profound:

DIVERSIFICATION OF ASSETS OFFERS INVESTORS BOTH PROTECTION AND PROFIT POTENTIAL.

I said it during the S&L Bank crisis of the '80s... during the stock market crash of '87... during the Gulf War crisis... during the Asian financial crisis... during the stock market bull of '91 through March 2000... during the Y2K scare... and during the dot-com meltdown from March 2000 through March 2001.

I have warned Americans again and again that debt and credit are like drugs - the more you rely on them, the more you need them to survive. But at what cost?

Alan Greenspan recently injected a surprise dose of credit and debt into the markets by dropping interest rates one-half percent right smack in the middle of the trading day. The result was a market rush that lasted a whopping 24 hours - then came the sobering reality that (this time) the credit drug alone may not revive the markets.

I was recently listening to a popular radio program who polled its listeners to find out if 2001 would bring a bull market or bear market. Some callers expressed fear and others faith, but it seemed everyone was still looking for the "inside scoop."

The host's advice to one caller asking whether to sell or hold stocks; "Sell, then reposition assets into something that you have long-term confidence in." My answer to the question: bull or bear?

IT DOES NOT MATTER... IF you are diversified!

Thankfully the pendulum has started to swing in a new direction - toward conservation, debt reduction, savings and tangible assets.

Most Americans are unaware of how well high-quality collectible coins and other tangible assets have performed historically due to preoccupation with the recent tech rush.

For example, last year U.S. Gold Commemorative coins outperformed almost every stock index and mutual fund - with a slow but consistent growth - amounting to as much as one percent per week. Yet virtually no one has heard about it except a few viewers of CNN's *Business Unusual* or PAX *MoneyWatch* when I was a guest.

In Greenspan We Trust?

Alan Greenspan and former President Bill Clinton are frequently given credit for the stock market boom of the late '90s, but the truth is they had very little to do with it. The real credit should go to the brave corporations who cut costs and downsized in the mid to late '80s - creating an atmosphere conducive to higher profits and future growth.

Another key factor of the '90s bull market was thriving consumption by consumers who abandoned savings and instead speculated by living and investing on borrowed money. And a third major factor was the promise of increased productivity resulting from technology and the Internet.

As a result, we have been programmed to believe the stock market is synonymous with the economy during the last decade. Not so. The economy is much broader than stock market speculation.

The real economy is based on what's happening on Main Street, which means that profits and earnings are the keys to sustainable growth - a painful truth that Wall Street is learning in 2001 (along with 50 percent of Americans holding stocks).

Bill Bonner's *Daily Reckoning* explains why the stock market must correct in 2001:

> *Demand for credit is saturated. Confident of an era of permanent prosperity, protected by Greenspan, consumers cut savings rates from nine percent in 1991 down to nearly zero in the year 2000. Household debt rose to 13.7 percent - near an all-time record.*
>
> *Life, as we have come to enjoy it, has been one long continuous boom. And the economy has enjoyed its longest continuous period of growth in nearly 40 years. As a result, too many people owe far too much money - and have too little to show for it. You don't solve these problems with lower interest rates. You solve them with bankruptcy, defaults, work-outs, cutbacks and real savings.*
>
> *Investment and business strategies rely on Mr. Greenspan to such an extent that he has come to assume a godlike stature. Not long ago,* Fortune *magazine ran a cover story called "In Greenspan We Trust."*

Will Alan Greenspan succeed in keeping the bull market alive? Will he manage the soft landing that he hopes for? I'm not so sure. More than one million bankruptcies are expected in 2001 alone - without forecasting a major bear market or recession.

So I ask, in whom can you trust if not Greenspan? Wall Street gurus? They're all playing the blame game for their bad market calls over the last two years. Recently even Congress has called Wall Street analysts into accountability for their "unbiased" forecasts.

For two decades I have suggested that investors learn to trust in historically sound economic principles - such as getting out of debt, increasing savings and diversifying into tangible assets - rather than trusting the world's wisdom.

2001 appears to be a year of financial reckoning and now is the time to invest in liberty and independence from Wall Street speculation. The best hope Main Street (and Wall Street) has in 2001 is a major tax cut led by George W. Bush - not lower interest rates. Alan Greenspan's effort to slather more liquidity on Main Street will not help the 90 percent of consumers who are already tapped out. Ditto for most corporations. They need to find more profits, so they'll need to cut expenses, which means downsizing, lay-offs and very possibly a recession in the economy.

The solution for confused investors is the same in 2001 as it has always been. **Diversify your assets, "...for you do not know what disaster may befall the land."** (Eccl. 11:1-2)

To ride the next economic wave safely into shore means getting yourself up to speed with what is really going on in the market and then diversifying a small portion of your portfolio into tangible assets like U.S. rare coins for long-term growth. Then, next time someone asks you, "Bull or bear?" You can honestly say, "Who cares, I'm covered!"

The Big Picture Challenge

During the last 20 years, I've tried to pour my heart, soul and mind into millions of investors, by announcing that true wealth is built on more than just money - it's built on peace of mind, time-tested principles and strong relationships.

Toward that end, I even produced a CD/booklet entitled "The Big Picture" to help my readers and listeners get up to speed and plan now for the next social, political and economic wave of change.

From my perspective, the bigger picture looks beyond the world as it is, to see the trends that are shaping the world to come. A world system that will increasingly become more volatile.

Thankfully, rays of hope are on the horizon as a fresh group of leaders begin to tackle the tough issues behind the news. America desperately needs more brave, self-governing leaders that embrace humility and morality - rather than mocking it.

Our nation and communities need more men and women of understanding and wisdom, who are less concerned with personal advancement. These new leaders will offer hope to our generation... and the next.

But, the next financial wave will bring with it a heavy undertow called debt. Like a deceptively strong underwater current, the undertow of debt can overpower anyone - including the strongest swimmers - unless they have a life jacket.

Throughout the last two millennia gold has consistently served as an economic life jacket. Smart investors are already reducing debt and diversifying assets to include tangibles - like investment-grade, historic gold and silver coins.

6.2 Economic Wild Cards
What's Going on Here?

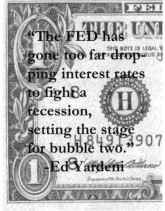

Stocks, energy, debt, deficits, inflation, money supply and gold were all up in the spring of 2001.

What's going down? Interest rates, savings rates, consumer confidence and national productivity.

"The FED has gone too far dropping interest rates to fight a recession, setting the stage for bubble two."
-Ed Yardeni

Is it any wonder that economists, media pundits and the public are all so confused?

I will now touch on just a few new problems that could become far worse than the $4 trillion blood bath we saw in the NASDAQ last year. Any one of the following wild cards could easily trump Greenspan's soft-landing approach and even perhaps develop into the worst economic event in world history.

Ed Yardeni, chief strategist for Deutsche Banc recently told CNNfn, "The Fed is setting the stage for bubble two."

My goal is that this information will help you avoid this emerging crisis and perhaps even prosper by having a winning strategy regardless of the final short-term economic outcome.

The Blame Game

In the face of confusing economic indicators America has discovered a new game - the blame game. It seems Americans now want to pin all their troubles on a person or a group of people, regardless of whether the facts support it.

The lawyers love it.

And why not? For eight long years we witnessed the Clinton administration elevate the blame game to a new art form. Now it appears that the public has developed a ferocious appetite for finding someone to blame for all our woes - politically, socially and economically.

In 2000, the almighty U.S. stock market slowly unraveled and everyone from politicians to the average guy wanted someone to blame for the drop.

We first looked to Alan Greenspan's Federal Reserve. The public outcry was a popular blame-shifting mantra, "How could the financial genius Alan Greenspan allow such a loss of value occur? He should have cut interest rates sooner!"

Then some turned the blame toward then-presidential candidate George W. Bush, claiming his frank comments of concern about the faltering stock market and economy caused it to happen - as though talking about economic realities creates a self-fulfilling prophecy.

Next Americans began to blame the market downfall on the leaders of the tech boom, saying: "How could they waste so much money building their castles in the air?" As if someone had held a gun to their heads forcing them to follow the media pundits perpetual slogan, "Buy tech!"

I can't remember when any network program has ever mentioned that it may have been the greed of the average investor for long-term, double-digit returns that contributed and fueled the market excesses. Perhaps I missed it. Did you see it?

Then there's the ever-popular blaming of Wall Street analysts. *Fortune* magazine's cover story, "Where Mary Meeker Went Wrong" (5/14/01) said: "Meeker came to see herself not only as an analyst but as a player - a power broker, a dealmaker, a force to be reckoned with. Meeker was the unques-

tioned diva of the Internet Age. Tech companies begged her to cover them. Morgan Stanley paid her an eye-popping $15 million in 1999. That was then. Today Meeker has become something else entirely. Though she was Queen of the Bubble, Meeker hardly reigned alone."

Prior to the tech wreck, inexperienced stock market investors just couldn't conceive that corporations would not be able to turn in quarter-after-quarter of profits to support the excessive prices of their stocks. But by 2001 betrayed investors were looking to blame.

The Energy Blame Game

In the spring of 2001 we also faced a new energy crisis in America. Gasoline prices hit more than $2 per gallon in many cities and were projected to possibly hit $3 by the end of the summer. Yet incredibly, most of our energy seems to be spent on pointing the finger of blame rather than discussing viable solutions.

Predictably the blame for our energy crisis was first aimed at President Bush and his "oil buddies," instead of the lack of attention this problem received from Bill Clinton and former Energy Secretary Bill Richardson. You remember him - he was blamed when secrets were leaked to the Chinese about the Los Alamos Laboratory on his watch.

Remember in 1999, when Clinton released the strategic oil stockpile to ease prices? The crude oil had to be sent overseas to be refined into gas and then sent back to the U.S. because our refineries were at full capacity. We must now find someone new to blame!

Perhaps we should blame the automobile manufacturers for building those gas-guzzling SUVs that have become so popular in America. Few have considered that part of the blame should be upon every-one of us that is using more gas or refuses to consider car pooling. Let's not forget the overblown environmental worries that have created a political climate that deterred us from building any new oil refineries for more than 15 years in California.

California is the biggest example of the energy blame game. It is Governor Gray Davis's fault, right? He is the one to blame - or is it those money-grubbing utility companies that are sucking the public dry by overcharging?

Breaking the Law of Supply & Demand

The California energy crisis is a perfect example of what happens when the free-market law of supply and demand is overridden by government regulators.

According to Hans Senhold, noted free-market economist, "The supply-and-demand principle points to three ways of creating an energy crisis. One, legislators and regulators fix energy rates that do not allow for rent, labor and profit to bring it thither. In modern terminology, they set energy prices below market prices. Two, legis-lators and regulators do not set rates but prevent producers from meeting a growing demand through stringent emissions and zoning rules, which causes the rates to soar until there will be a "rate crisis." Third, legislators and regulators may do both, fix rates below the market and prevent the supply from adjusting to the demand, which is bound to create a double-prong crisis.

"California political leaders chose the double-prong approach. In

1996 California legislators unanimously passed a 67-page electricity-restructuring law. They called it "deregulation" by which, in typical political duplicity, they meant re-regulation.

"It contained something for all the different players. The number of regulators was increased with the addition of two new quasi-governmental organizations - the Power Exchange, which controls all transactions between utilities and electricity generators, and the Independent System Operator, which operates the electricity grid, purchases the needed power and charges the utilities.

"Consumer groups got an immediate 10 percent rate cut and price caps at roughly the 1995 level. Powerful environmental groups were reassured of stringent environmental rules and zoning restrictions. The utilities, finally, were given strong financial incentives to sell their fossil fuel-fired power plants. They subsequently reduced the self-generated power from 72 percent to just 20 percent, purchasing the balance on the market.

The government's solution? Mr. Senhold concludes, "The legislature created another power authority: the California Consumer Power and Conservation Financing Authority which may issue bonds to build state power plants. Moreover, it is preparing legislation authorizing the State to issue more bonds to buy the transmission lines from the utilities. If they refuse to sell, the governor threatened to use the power of eminent domain. In short, the State is taking over the electricity industry."

So the blame game is on a roll in California, but the entire country will pay the price as inflation continues to increase the cost of living. Face facts - big business and the government can avoid inflation's effects by increasing prices or decreasing services. The average citizen cannot unless they are willing to think outside the box created by today's inflationary money system.

Inflationary Trends
Expecting the Inevitable

Inflationary pressures are rapidly changing. Some investments will soar, while many more plunge. And anyone who bets on a continually rising market in which everybody profits, is destined for failure.

Inflation is the one wild card that can whip investment markets into a frenzy, cut your true wealth by half in just a few years and completely derail an otherwise sound investment strategy.

In the late '60s inflationary surprises ripped through the stock market, dropping the Dow some 35 percent. In the 1970s stocks suffered through a tortuous two-year bear market as inflation spiraled. In the '80s Paul Volcker nearly plunged the U.S. into a 1920s-style depression when he miscalculated the effect of "runaway inflation."

Now after years of relative calm on the inflation front, the risk of grievous error is just as strong as at any point in history.

USA Today (6/11/01) reported that "Inflation could outpace savings for the first time in seven years." They went on to say that holders of CDs and Money Market funds yielding 3.5 to 4 percent are now facing a guaranteed loss due to rising inflation.

People who know me know that "doom and gloom" has never been my style. However, I do feel compelled to warn you about the risks looming ahead of you. Inflation poses an immediate danger, but also a long-term threat - especially as you reflect back on how easily you could have been protected by taking a few simple steps now.

Why do most investors overlook inflationary trends? Until recently productivity has rocketed ahead on the back of revolutionary technology that lets one worker do the job of 10. And billions of eager new capitalists have turned the bustling economies of Asia, Eastern Europe and Latin America into a competitive force like never before. But in 2001 it appears that inflation could skyrocket. Why?

Greenspan is Cornered

Faced with the onset of a business slowdown, falling stock prices, and the possibility of a recession, starting in January 2001 the Fed did just what everyone expected - they reduced the price of money. In fact, it did so more aggressively than any other time in history - dropping rates from 6.5 percent to 4 percent.

Chris Low, chief economist at First Tennessee Capital Markets confirms the inflationary implications: "Greenspan has become completely reactive. Either we'll have a recession, and he'll get blamed for being slow to ease, or he'll engineer a recovery, which will be accompanied by an inflation problem. Take your pick. Steering a middle course would be a miracle."

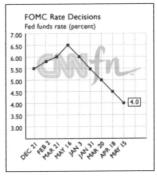

Caught between inflation and recession - there is no doubt about which way Mr. Greenspan will steer the economy. He would rather risk the whirlpool of escalating prices than risk being eaten alive by recession.

"The Fed chairman has few tools in his workshop. He can regulate the quantity of money, or its price," says *International Speculator*'s Jim Grant, "but not both at once. For the past 20 years, the Fed has chosen to regulate the price, i.e., the federal funds rate. The quantity, i.e., the monetary base, is what it doesn't control."

The Fed is doing what it said it would not do, easing policy in the teeth of rising inflation rates. Why? Because they believed inflation would not be a problem due to increased productivity.

Rising levels of productivity - providing more and more goods and services per unit of input - were supposed to offset increases in the supply of money. But what if U.S. productivity continues dropping, as it has for the last three quarters?

Productivity has been a key factor in the economy's last 10 years of growth - the longest expansion in U.S. history. As workers produce more per hour, companies can sell more, helping profits while keeping workers' wages in check and limiting inflation pressures.

Many economists have credited productivity gains in large part to businesses' investments in computers and other technology, but some fear that the productivity boom is over, at least for now.

Slowing productivity and rising labor costs raise the specter of inflation. Clearly inflationary pressure is building and lower interest rates offer Americans no guarantee that the money supply will not explode in the days ahead - causing double-digit inflation to crush any rally in the stock and bond markets.

Bill Bonner's *Daily Reckoning* confirms that the Fed cannot control the inflationary trend by dropping interest rates indefinitely:

> *"A lower fed funds rate - the rate that the Fed charges member banks to borrow money - allows the banks to lend at lower rates too. In a deflationary downturn, money becomes hard to get. People lose jobs, the values of stocks and other investments fall, sales and profits decline... and there are still big debts to pay. Lowering the price of money may have some effect, but not necessarily the desired one."*

> *"People who are out of work and owe too much money do not make good candidates for additional lending. Nor do businesses want to borrow for capital improvements when the price of capital itself is in decline."*

> *"In Japan, the price of borrowed money has been reduced to zero - with little effect. In America, Fed policies were so accommodating and the bull market so enriching that the real cost of borrowed funds could be said to have been below zero for many years. An investor could borrow*

at 8 percent on a home equity line, and invest the money in the stock market for an 18 percent return. The net cost of money - minus 10 percent."

"But, if you put the price of money low enough, it is argued, people will see that it is worthless and want to get rid of it quickly. The government can always crank up the printing presses. But this is a little like ending a war by committing suicide."

To protect your assets against inflation you must have your assets diversified into many areas - including tangibles like coins, gold and land. Yet this is a strategy that scarcely one network talking head will ever suggest because tangible assets are in competition with the debt and equity asset markets.

A Golden Spike

Forget about stocks for a moment, the attention now is shifting toward the gold market. Gold spiked $13.80 per ounce on May 18, 2001 - about 5 percent. This surprisingly strong advance was the long-slumbering precious metal's biggest gain in 15 months. Why?

Let's add it up: the money supply is soaring... consumer prices are rising... and the Fed is cutting rates. No matter how you slice it, inflation is on the ascent.

In his frantic effort to avoid a recession Alan Greenspan has opened the floodgate of money - increasing the total money supply by a staggering 28 percent in the first three months of 2001.

According to Richard Russell's *Dow Theory Letter*, "Gold loves it. Throw an extra trillion dollars into the mix and it's got to stir up something. That something is housing, medical bills, restaurant prices, sports tickets, utility bills, and well - most of the items you buy in daily life."

In fact, almost all measures of inflation are accelerating at a time when the Fed is stepping on the accelerator to stimulate demand.

Bloomberg.com said, "The Fed has lost sight of its long-run objective of price stability. Rising inflation would be one thing if the raw material of inflation - money - was signaling disinflation ahead. Instead, April will be the fifth consecutive month to see double-digit annualized increases in the broad monetary aggregates, M2 and M3."

What will this inflationary trend mean for a skittish stock market, and how will increasing bankruptcies affect the banking system?

Good Times = Bad Loans

Years ago Americans watched in horror as the Savings and Loans precariously hung by a thread on the precipice of failure until the Resolution Trust Corp. was formed to bail them out at huge taxpayer expense.

In 1987 the banks single-handedly saved the U.S. stock market by pouring billions of dollars of liquidity to stop the bleeding of a 20 percent "correction."

In the years since then the equity markets appeared to be the only game in the investment world. So much so, that the average person ignored the distinction between investment and savings and invested his life savings into a stock market which, by historical measure, was grossly overvalued.

During the prosperous '90s, many bank loans were made; everything from margin loans on inflated stock values to 150 percent mortgages on our homes, which fueled the economy and the stock market's growth.

With billions of dollars of freshly borrowed money, the world looked like a wonderful place. People were so drunk with the so-called "new era" that most even believed they could eat unlimited amounts of candy without ever gaining a pound or getting a cavity. There is an old saying in the investment community: "Good loans are made in bad times and bad loans are made in good times," made famous by Fed Chairman Alan Greenspan.

This statement makes sense. When times are bad, a bank or any lender for that matter will only lend money to those that they know have the ability to pay the money back. However, during good times a lender will loan to anybody who has even a remote chance of paying back the loan. The logic is that the lender has enough profitable loans to offset any losses.

Keep in mind we've just gone through one of the best ten-year periods of good times we have seen in the nation's history. Thus if the old saying is consistent, as it has been in the past, we have got a lot of bad loans out there that will surface soon.

Unlimited Money Supply = Zero Savings

If $A+B=C$, then we are in bigger trouble than we know. Just as the recent rash of school shootings illustrates a much deeper problem in society, so the rampant increase in the U.S. money supply should be sending us a clear economic distress signal.

As mentioned, the Fed has recently increased the money supply at a rate of 28 percent - an unprecedented rate. They are creating more liquidity than was made available during the crash of 1987 or in preparing for Y2K. They are also keeping interest rates low to make that money very attractive to borrow. The same magical formula that worked so well in 1987 appears to have little if any effect on the market or the economy. Why?

My opinion is that most people in America - and the world for that matter - are tapped out and scared. Both *Time* and *Newsweek* devoted several covers in spring 2001 to discussing the growing fear in America. Why all the fear suddenly?

I think most people are scared to death that the explosive stock growth (read: irrational exuberance) that we experienced for the last decade is gone now and it may not come back soon. America is afraid that we may just have to tighten our belts and live within our means and stop buying everything under the sun whether we can afford it or not.

A Confidence Crisis

Here's a troubling thought: Just because the Fed rate cuts are credited for rescuing the NASDAQ, thereby preventing Cisco shares from trading in single digits, it does not mean that the Fed has also saved us from recession.

It's quite possible that both the consumer and corporate America will avail themselves of easy credit, but will they use that credit to consume or invest? Not necessarily. Instead I think most will maintain their lifestyle and existing debt load rather than cutting back.

Economists and politicians know Americans are over their heads in debt. Lawmakers changed the bankruptcy laws in early 2001 to afford the credit card companies more protection. Incidentally, the credit card departments are the most profitable operation in a bank.

Politicians even pushed a tax cut through the House and Senate because they know people need money and that it irks the public to see budget surpluses sitting in government coffers instead of being returned to taxpayers.

Politicians are savvy enough to know that if the banks did falter, the whole charade is over. The banks and the Fed money system work completely on confidence - if that confidence fails, the entire system could collapse virtually overnight.

I'm not saying the banking system is collapsing, that would not be politically expedient. What I am suggesting however could be more devastating than a banking collapse. Keep in mind, in 1929 when the banks did collapse, the citizens knew the reality and could defend their savings and investments accordingly. It won't be that way this time.

A "too big to fail" philosophy exists today, and the accepted means of keeping the system intact is to inflate the money supply ad infinitum. This increases the liquidity to big businesses, while contracting liquidity to the average guy - allowing the effects of inflation to be shouldered by the public through higher prices, not by government or big business.

America's economic and banking systems are both in trouble and it is probably going to get worse before it gets better. And when the markets and the masses cry out to their savior Alan Greenspan for the answers, it will fall on deaf ears.

Mr. Greenspan knows that some pain has to come in order for the system to come back to some sense of reality. He warned us about "irrational exuberance." He knew that P/E ratios could not stay at 200-300. He knew the banks could not keep making outrageous profits on the ignorance of the average person. Can anyone actually justify paying 18-21 percent on credit card debt? Yet millions of people do it every single day without giving it a second thought.

Ending The Blame Game

The fact is that we have no one to blame for most of these problems except ourselves - which means taking personal responsibility for the world around us. America's Founding Fathers envisioned a nation of responsible, self-governing individuals and communities.

Instead of blaming, America's founders took responsibility. They willingly shed their blood and gave up personal fortunes to secure our future and the personal freedoms that we hold dear. Yet today the majority still want someone else to be in charge - to be the responsible party.

American statesmen like James Madison assumed
that future generations would choose wise leaders to
represent them and govern by putting the common
good of the country ahead of their personal interest.
They deduced that if elected officials violated our
trust they would be removed from office and never
re-elected. I guess they did not figure on such a high
level of public complacency.

James Madison

No, the Founding Fathers never foresaw that the population would
become so caught up in the pursuit of wealth that they would vote
their pocketbooks rather than their conscience. They thought "We
the People" would always do the right thing because they always
sought to do the right thing. History seems to prove that was a bad
assumption.

Sadly, most Americans cannot even define money today, so when it
disappears very quickly they are frustrated that their luck has run out
and must blame someone. Perhaps the best way to end the blame
game is to understand the "game" well before playing it at all.

Work, Risk & Luck

There are only three ways to create wealth...
1) to work
2) to take risks
3) to rely on luck

Yet today the masses want wealth without risk, without much work
and with good luck following them around like a shadow in the
afternoon. Much of this is due to the rise of a lottery mentality,
which is promoted at almost every level today.

There was a time in America when great invention and marvelous
discoveries were sought after for the primary purpose of improving
mankind. Most of America's great inventors cared more about making the world a better place than about the wealth that might follow

- men (and women) like Edison, Salk, Franklin and Pasteur.

There was also a time when savings was a sign of maturity - when debt was shunned like the plague. But not today, it is just the opposite - and now Americans can't figure out why they feel like they are walking up a down escalator, financially. I believe we live in the greatest country of opportunity in the world, but we must keep things in proper prospective.

Yet when someone or some group in society doesn't have wealth our first reaction is often to blame someone or something else. Rarely do we consider that the reason someone doesn't have a job could be that they are just plain lazy or not skilled to do the job right. Or if you didn't make as much money as your friend on an investment, could it be he was willing to take more risk than you?

Could it be that we take inexpensive fuel for granted? Could it be we will be forced to discipline ourselves to drive less? If we have extended rolling blackouts in California, could the reason be that we don't hold our politicians and leaders responsible?

Could our country be in the present moral crisis because we elected, then re-elected, an immoral man to the White House? You see the problem with the blame game is that even if you do pin the blame tail on the right donkey it doesn't fix anything unless you do something with the donkey. What we need to do is to analyze the problem and see how we can fix it. Not the politicians - us! Until we begin to take responsibility and stop playing the blame game, I expect growing volatility in the financial markets and gold prices.

The Untold Story of Gold

Throughout history, civilization has always had a tremendous appetite for gold. As a symbol of beauty and wealth, gold has been desired by the New World explorers to the new Eurodollar bankers. Whether hunting for lost treasures or exploring new frontiers, mankind has made the ultimate sacrifice in its pursuit of the yellow metal.

Given the rich history of gold, I marvel at the countless articles written in the 90s discussing gold's diminished role as a monetary asset. While nations once waged wars to get their share of the precious metal, recently central bankers of Belgium, the Netherlands and Australia have sold their gold reserves off at astonishingly low prices.

It was truly astounding that Australia would sell millions of ounces of gold, driving the price down and diminishing the value of their in-ground gold reserves in order to invest the proceeds into fixed income assets.

Leave it to politicians and bankers to violate two fundamental economic rules - value and timing.

Even more damaging to the gold price during the last few years has been the practice of gold loans by the large bullion banks. It had been a common practice for these banks, which had large positions of gold, to loan the metal to gold mining companies at low interest rates. The mining companies could repay the loans at the time of their production. Rather than speculation, this is referred to as hedging in the industry.

An analogy would be a farmer who harvests corn in September, but sells the contract in May when the prices are high so that he is guaranteed a nice profit for his future harvest. The bankers were making interest on a non-working asset - gold - and they decided to expand their gold loans. Large brokerage companies and hedge funds took advantage of low interest gold loans at 1 to 2 percent - that is where the hedging ended and the speculation began.

Murphy's Law

In 1998, Mr. Bill Murphy, a former professional football player and long-time commodities specialist, formed the Gold Anti-
Trust Action Committee (GATA) to look
into allegations of collusion and manipulation in the gold markets.

Bill Murphy (left) with
South African King at May
2001 Durban conference.

GATA hired Berger and Montague, a prominent anti-trust law firm,
and gathered the support of the large gold mining companies.
Together they went to the international press and the U.S. Congress
to report on the reckless gold loans of the bullion banks.

Apparently, the hedge funds and brokerages that borrowed the gold
at 1 to 2 percent sold the metal in the futures market and reinvested
the proceeds into significantly higher-yielding investment vehicles.

As long as there was a continuous new supply of gold being loaned
and sold, the price would continue dropping. It was then profitable
for the funds to cover their short gold positions and buy back the
metal at a lower price. If one sells an item high and buys it back low
that is a guaranteed profit in the commodities futures market.

GATA and Murphy were dumfounded as to why the price of gold
continued to drop while the prices of basic commodities (e.g. oil)
were rising during the early part of 1999.

The last outrage occurred just when gold was rising in May to
approximately $290/oz. Out of the blue, the Bank of England
announced it was selling half its gold supply. However, before the
first sale of 25 tons, the price plummeted to $30/oz. The British had
lost millions of pounds prior to the sale!

Imagine a major mutual fund announcing that they plan to sell millions of shares of IBM and the stock then plummets before the sale
is completed, causing the shareholders to lose millions. While the
mutual fund shareholders would quickly reorganize management,

the British had no recourse; no one took responsibility when their gold reserves as well as their pound currency lost value. GATA and Murphy believe that insiders at several major U.S. brokerage houses were colluding with insiders in England to drive the price down, believing they were short hundreds of tons of gold.

GATA's main concern, as expressed to the members of the U.S. Congress, is that the international banks recklessly loaned excessive amounts of gold. An analogy would be the reckless loans made by savings and loans executives in the early '80s, which left the government and taxpayers holding the bill of billions in "bad loans."

GATA fears that up to 10,000 tons (not ounces) may have been loaned. With most of the borrowed gold having been sold and with a supply demand imbalance occurring, how will this gold be replaced?

If the price of gold escalates, these hedge funds must quickly cover their short positions, buy back the gold or risk going bankrupt. Is it possible that large financial institutions as well as large international hedge funds are at risk?

If Douglas Casey is right and we are now entering a "bull supercycle" in gold, the fireworks have already begun. Maybe the European bankers recognized their plight when they announced on September 26, 1999, a restraint in further gold loans, a tightening of credit on gold loans and a limit of gold sales to 400 tons per year. Hence, the price of gold exploded from $252 to more than $320 within just a two-week period.

Goldgate: Gold Derivative Banking Crisis

The following is "An Open Letter to Senate and House Banking Committee Members" by Bill Murphy, founder, GATA.

Read it carefully to understand why it is only a matter of time now before the truth is exposed and gold prices skyrocket.

Extensive research has led the Gold Anti-Trust Action Committee to the conclusion that the gold market is being recklessly manipulated and now poses a serious risk to the international financial system.

Annual gold demand, currently at record levels, exceeds the mine and scrap gold supply by more than 1,500 tons. In the Washington Agreement of September 26, 1999, 15 European central banks announced that they were capping their lending of gold and would limit their official sales of gold to 400 tons per year for the next five years.

Some major gold producers have reduced their forward sales, and speculators have reduced their borrowed gold selling. Commodity prices and wages are rising. Yet the price of gold has declined steadily. With demand so much greater than supply, the price of gold should be rising sharply.

According to the Office of the Controller of the Currency, the notional value of off-balance-sheet gold derivatives on the books of U.S. commercial banks exceeds $87 billion, which is greater than total U.S. official gold reserves of approximately 8,140 metric tons.

Gold derivatives surged from $63.4 billion in the third quarter of 1999 to $87.6 billion in the fourth quarter, after the Washington Agreement was announced. The notional amount of off-balance-sheet gold derivative contracts on the books of Morgan Guaranty Trust Co. went from $18.36 billion to $38.1 billion in the last six months of 1999.

Veneroso Associates estimates that the private- and official-sector gold loans stood at 9,000 to 10,000 tons at the end of 1999. Most of these loans represent gold that has been sold in the form of jewelry and cannot be retrieved. Mine supply of gold for all of 1999, according to trade sources, was only 2,579 tons. Thus the gold loans are far too big to be repaid in a short time. The swift $84 rise in the gold price following the Washington Agreement caused a panic among bullion bankers. But that was only a warning of what is to come.

Federal Reserve Chairman Alan Greenspan and Treasury Secretary Lawrence Summers, responding to GATA's inquiries through members of Congress, have denied any direct involvement in the gold market by the Fed and the Treasury Department. But they have declined to address whether the Exchange Stabilization Fund, which is under the control of the Treasury Secretary, is being used to manipulate the price of gold.

Several prominent New York bullion banks, particularly Goldman Sachs, from which the immediate past treasury secretary, Robert Rubin came to the Treasury Department, have moved to suppress the price of gold every time it has rallied during the last year.

The Gold Anti-Trust Action Committee believes that U.S. government officials and these bullion banks have induced other governments to add gold supply to the physical market in recent years to suppress the price. Britain's National Accounting Office is now investigating the Bank of England's decision to sell off more than half its gold. Contrary to proper accounting practice, reductions in gold in the earmarked accounts of foreign governments at the New York Federal Reserve Bank are being listed by the Commerce Department as the export of non-monetary gold. These exports from the Fed occur upon rallies in the gold price.

Why would anyone want to suppress the price of gold?

First, suppressing the price of gold has made it a cheap source of capital for New York bullion banks, which borrow it for as little as 1 percent of its value per year. Gold is borrowed from central banks and sold, and the proceeds are invested in the financial markets in securities that have much greater rates of return. As long as the price of gold remains low, this "gold carry trade" is a financial bonanza to a privileged few at the expense of the many, including the gold-producing countries, most of which are poor. If the price of gold were allowed to rise, the effective interest rate on gold loans would become prohibitive.

Second, suppressing the price of gold gives a false impression of the U.S. dollar's strength as an international reserve asset and a false reading of inflation in the United States.

Too much gold is being consumed at too cheap a price. Massive amounts of derivatives are being used to suppress the gold price. If this situation is not corrected soon, there will be a gold derivative credit and default crisis of epic proportions that will threaten the solvency of the largest international banks and the world standing of the dollar.

As you are aware, a 90-page document of our extraordinary findings was personally delivered to your offices last Thursday.

The Gold Anti-Trust Action Committee requests that a full and complete investigation be launched into this matter as soon as possible.

The longer the gold price is artificially held down, the bigger the eventual banking crisis.

Gold Anti-Trust Action Committee, Inc.
Bill Murphy, Chairman, LePatron@LeMetropoleCafe.com
Chris Powell, Secretary / Treasurer, GATAComm@aol.com
Ethan B. Stroud, Attorney at law, formerly Justice Department, Treasury Department
John R. Feather, Attorney at law, formerly legal staff, Federal Reserve Bank
Suite 1203, 4718 Cole Avenue, Dallas, Texas 75205
(214) 522-3411 phone (214) 522-4432 fax - www.gata.org

GATA Delegation Progress (5/2000)
By Bill Murphy, GATA

On Wednesday, May 10, 2000 at 11:30, the Gold Anti-Trust Action Committee consisting of Chris Powell (newspaper editor), Reginald Howe (lawyer), Frank Veneroso (macro-economist), a State Senator and I met with one of the most powerful politicians in Washington. It was only going to be a 15-minute meeting - it lasted an hour.

At the end of the meeting, we were excused from the room for several minutes. When the people we met with returned, we were told that they were going to try to set up a meeting with another influential politician at 2 p.m., but that we would have to call at 1:30 to confirm.

We were stunned to learn at 1:30 that this politician had said, "I am aware of the issue," and that he wanted to meet with the GATA delegation. The meeting took place and six members of his staff also attended. What was most remarkable is that this politician left the floor of Congress to attend our hastily arranged, unscheduled meeting.

This politician asked many questions and was very focused on what we had to say. So much so, that he was annoyed when a staff member left to deal with some other pending issue, saying that this was more important. He told us he had read our biographies before coming to the meeting and was a bit taken aback when he was handed the "Gold Derivative Banking Crisis" document with his name and state on it.

This knowledgeable politician said that he and his staff would look into our contentions and suggested that we might meet again. After this very intense one-hour meeting, he returned to Congress, which was in session. From there, we went on to meet with Dr. John Silvia, the Chief Economist of the Senate Banking Committee. I could tell he had spent some time on our presentation because he had highlighted material that I had sent to him.

Frank, Reg and Chris did a terrific job (as they did in all the meetings) explaining what we have learned through our extensive research. That meeting also lasted an hour and Dr. Silvia took copious notes.

Yesterday, I passed out 88 of the documents to the staff of all the Senators and Representatives on the banking committees. They were told to look for an open letter to all of them in Monday's Roll Call. For the Senate I went to the Dirksen, Russell and Hart buildings. For the House I went to the Rayburn, Longworth and Cannon buildings. It took me the entire day, but was well worth it. Congressman Lee Terry of Nebraska could not have been nicer and said he would read the document on his way back to his native state this weekend.

My last stop was the Rayburn Building and I smiled as I went by The Gold Room. It was the opinion of the entire Gold Anti-Trust Action Committee delegation that the trip was far more fruitful than any one of us dreamed possible. However, as we all know, that was just our first salvo. There is much to be done to win the day and we are already planning our next course of action.

When our adversaries realize how far we have come, we know that they will go all out to discredit us. If yesterday's meetings were any indication of making a serious impact on those who count in Washington, the other side has their work cut out for them.

The GATA Lawsuit
Fed, Treasury and Five Investment Houses Sued (12/2000)

A lawsuit filed in U.S. District Court in Boston on December 7, 2000 with the support of the Gold Anti-Trust Action Committee (GATA) accuses five investment houses, the Bank for International Settlements, and top officials of the U.S. Treasury Department and U.S. Federal Reserve Board of conspiring to suppress the price of gold.

The lawsuit charges the defendants with price fixing, securities fraud and breach of fiduciary duty. The U.S. government officials are also accused of exceeding their constitutional authority.

The lawsuit's plaintiff is Reginald H. Howe, a lawyer, gold market analyst, consultant to GATA and a shareholder of the Bank for International Settlements.

The BIS's plan to cancel those of its shares in private hands so that the bank might become owned entirely by member central banks is at the center of the lawsuit.

The suit alleges that the BIS proposes to pay its private shareholders substantially less than fair value for their shares. The suit also claims that the BIS, owner of a substantial amount of gold, has been at the center of a scheme with central banks and the investment house defendants to coordinate the sale and leasing of gold and the sale of gold derivatives to keep the price of gold low and thereby disguise inflation and weakness in the U.S. dollar, as well as to prevent losses on gold short positions held by certain banks.

Besides the BIS, the defendants include: Alan Greenspan, chairman of the Federal Reserve Board; William J. McDonough, president of the Federal Reserve Bank of New York; Lawrence H. Summers, former secretary of the Treasury; and J.P. Morgan & Co. Inc., Chase Manhattan Corp., Citigroup Inc., Goldman Sachs Group Inc., and Deutsche Bank AG.

The text of the lawsuit is posted online at: www.gata.org.

Indeed, the world is watching the price of gold and I am so thankful for the work that GATA has done calling our financial leaders into accountability. As of the publishing of this book the GATA lawsuit is pending a decision from the judge to proceed into the "discovery" stage. If this is granted, Mr. Greenspan, the BIS and all of those named in the lawsuit will be forced to give sworn testimony. Will the truth prevail? Time will tell.

The final chapter of *Rediscovering Gold in the 21st Century* is a transcript of an interview conducted by Pat Boone with myself and a few members of the Swiss America management team in May 2001. My hope is that this will answer any remaining questions and inspire you to rediscover gold for yourself in the days, weeks and years ahead.

7

The Swiss America Story

7.1   Rediscovering Gold with Pat Boone

Featuring interviews with Craig Smith, Earl Brown, Dr. Fred Goldstein and the Swiss America team. This chapter is a transcript of the compact disk.

PAT BOONE: Welcome to this special 2001 Market Update brought to you by Swiss America. The time again has come to reflect on the market trends of the last year and prospects for the next year with my favorite tangible asset brokerage, Swiss America.

Many years ago I discovered Swiss America and today I invite you to "rediscover" gold in the 21st century with the whole Swiss America team, led by Craig R. Smith. Swiss America is the only brokerage that I know of whose clients were prepared for the so-called stock market "tech wreck" of the last year.

So listen carefully as I ask Craig Smith and a few of his top brokers to give us their best advice for the year 2001 and beyond.

Craig, last year we recorded an interview entitled, "Investing Wisely in the 21st Century," and you said that you were concerned about the high-flying tech stocks. You said they were built on an inflated foundation and did not have unsustainable growth.

Well, one year later, 60 percent of the NASDAQ market value (or nearly $4 trillion) is now history - dragging the average investor's net worth down dramatically. I remember that you strongly suggested that listeners take the time to diversify a portion of their assets into tangibles, like U.S. rare coins. Sadly, most investors were blinded by the media hype on tech stocks and did not take your advice.

Nature loves symmetry. After the hectic storms of life comes the peaceful calm. Now that the dust has apparently settled on the tech stock market boom, what do you see ahead this year for our nation's economy and stock market?

CRAIG R. SMITH: I think we're probably going to have a very slow year 2001. I think there are a lot of factors involved. We've seen the slowdown in the pace of manufacturing, we've seen a tremendous amount of layoffs announced recently, as a matter of fact in the first three months of the year 2001 we saw over 200,000 job layoffs announced by major corporations like Motorola and many others.

If we had 1 to 2 percent growth rates I think we'd be satisfied. I don't think you're going to see the Fed quick to continue to cut rates, we might see another half point or maybe a full point during the year but I think the Fed is pretty much finished cutting rates. I think they realize you can't cut much more. They learned from the experience in Japan that you can't get rates to zero and have it create an economic boom. There has to be, if you will, a "norm" in there, you can't cut too much or raise too much.

I think the tax cut is going to have an impact on the U.S. economy. The quicker and the deeper that those tax cuts are, and the retroaction to January are going to be very essential to bringing some desperately needed capital into the marketplace, instead of borrowing it from banks like we've done in the past. That's one of the big problems, the underlying problems in our economy is that the banks are tapped out. The average American is tapped out. We've borrowed on everything from cars and homes and electronics, and credit cards, and we've even borrowed money in our stock accounts to purchase more stock. Those are going to have a negative effect on our economy.

Now that's the nation's economy, Pat, let's shift gears because a lot of times people talk about the stock market and the economy as one in the same and they're not. The stock market I'm not so optimistic about, as you know Pat, we've talked about it several times before. I've been pretty bullish on the stock market for the last 7-8 years because quite frankly there has been nothing to stop it - there has been an irrational exuberance and the momentum has been such that even when there was profits there and there wasn't reality there, investor momentum continued to propel the stock market higher and higher.

We've been warning our clients for several years now that unless earnings followed this speculation, that we'd see this type of pull-back and of course it's happening now: price-to-earnings ratios are coming back to more realistic levels. We believe that P/E should be no higher than 20 in tech companies, and for average, good old-fashioned companies should be somewhere around 8-16 times earnings.

And so the stock market I'm not so optimistic about, I think we're in what's called an L-shaped correction - it's not going to be a U-shaped or a V-shaped - you saw the stock market drop and now it's going to go sideways for probably the next 12-13 months. There'll be some rallies along the way, you'll see 200 to 300 day increases, you might see 1000 points during the week increase but I think they will be short-lived, I think they will be what are referred to as "sucker rallies" and I think the market will drop.

So I'm not very optimistic about the stock market in 2001, but I am going into 2002 - I think in 2002 we will see a major shakeout, but you know Pat if we look at the last 10 years of growth in the stock market and we apply that to today's levels, most of these companies are going to take at least seven years to get back what they've lost, as it relates to stock value. And in the case of some of these tech stocks it could take 25 years before they ever get back to some of these levels, I mean I think of some of the stocks like Iktomi or Akamai that were $270 a share and today they're $11 a share. It's going to be a long time, Pat, before you see these companies ever get back to where they were, if they ever do get back there.

So I'm neutral on the economy, I think the economy's going to be okay its going to grow at a rather slow pace Pat, I'm not so optimistic about the stock market, I think now is the time to take a defensive posture. I think there's nothing wrong with sitting on the sidelines in cash, I would buy only valued companies, I'd be looking at blue-chip companies that have a long-term track record of growth and profitability. Without growth and profitability, I think they're going to be very hard-pressed to see their stock values go up this year.

PAT: Craig, a few years ago I visited your offices in Phoenix, and was very impressed with your slogan on the wall which says: IF WE DON'T TAKE CARE OF THE CUSTOMER, SOMEONE ELSE WILL. Give us your company philosophy in a nutshell.

CRS: Pat, I'm glad that you brought that up. You know when you walk into our office the first sign you see is, "If we don't take care of the customer, someone else will."

And actually Pat, I stole that from the Marriott Corporation. I saw that in the Marriott, it was years ago, and I thought to myself, boy that's really appropriate for our company. And that is our philosophy as a company, Pat, our future as a company depends upon our clients being satisfied with the service, with the product, with the experience they have had at Swiss America.

And although we can't dictate the movements in the market, Pat, I mean, I wish I could be a "soothsayer" if you will, and be able to tell what the future holds. We can't predict whether the markets will go up or down, we use the best data we can. We can't control that.

But what we can control is the customer experience, when they walk through the doors of Swiss America, or speak with one of our representatives, or get one of our educational packets.

Pat, it's been my heart's desire for the last 20 years in this business of bringing people into the tangible arena, if you will, educated. I don't want people just doing it because it sounds good, I don't want people purchasing tangibles out of fear or out of lack of knowledge, I want them to do it with an education, with a foundation that says this makes sense for my portfolio and here are the reasons why.

And I believe, Pat, that if you look at great companies - and that's what I want to be seen as in the years to come - is a great company. They look at making the customer satisfied so that they have a customer 5, 10, 15, 20 years down the road.

Look I know many of my contemporaries are happy to do business one time with a client and never talk to them again. That's not our philosophy. When we get involved with a client, if you will, Pat, there's almost a marriage that's created. It's Swiss America and the client and the broker working together for a successful and pleasurable experience in the area of tangible investment. And that's something that I work on every single day.

I hold meetings several times a week with my people, we're constantly increasing the education levels of our people, we have people working overtime on the websites, keeping them updated with fresh information, new reports, new informational CDs, Pat, you know as well as I do we have done everything from video tapes to cassette tapes, to CDs to printed reports to radio shows to television interviews, CNBC, CNNfn, the list goes on and on. Trying to make sure that our clients are the most informed clients in the country when it comes to tangibles.

Resources

PAT: Over the years I have also been impressed with the level of teamwork at Swiss America. What is the secret of building a great brokerage team?

CRS: You know Pat, I don't know if there is such a thing as a secret to building a great brokerage team. I suppose what separates my brokers from anybody else is the philosophy that I have taught them what I really truly believe, and that is 'people don't care how much you know, until they know how much you care.' And I've told my brokers, and I've tried to beat it into them, instill it in them that they have to care about their clients.

You see, a client is not just another voice on the other end of the phone. It's not just a paycheck, it's not just somebody to do business with. That person's a human being, they have wives and husbands, sons and daughters, moms and dads, aunts and uncles. They have trials and tribulations, they have triumphs, they have good times and bad times.

What I try and teach my brokerage team is that if we truly have an interest in the client as a human being first, and then as a client, we have a better opportunity and a better way to give them what they need.

You know Pat, it's kind of like a doctor, sometimes a doctor doesn't give you what you want, they give you what you need. Now there are a lot of other brokerage firms like ours that will give you what you want. You call them up and say here is what I want, and it might not be the right thing but they give it to you anyway. I'd prefer to pass on business like that, than to give somebody business that they don't need.

I suppose that's been the secret, as you put it Pat, to our success in building a great brokerage team, has been meeting and caring about people. Again, Pat, I hate to be repetitious, but I am absolutely and thoroughly convinced that people don't care how much you know, until they know how much you care.

It really works in our communities, in our churches, in our civic organizations, when you care about people, then the business comes as a by-product of that. I tell my brokers, you should treat your clients like you would your mom or dad, or your brother or sister.

Years ago for a very short time I was in the car leasing business, I leased cars for a living, and I remember a fellow I worked with - he would lease everybody nothing but Cadillacs - boy Cadillac was the best car in the world - and yet he drove a Mercedes-Benz, and I used to say to him, Jeff, how can you be so high on Cadillacs when you drive a Mercedes-Benz? And his answer distressed me, it was because he could make more money selling Cadillacs than Mercedes-Benz because not everybody could afford a Mercedes-Benz. And I really saw a contradiction there, I would have rather seen Jeff say to people, Look you should buy a Mercedes, if you can't, a Cadillac's a great second car, or a great second choice.

I believe you've got to be honest with people, give them the whole story, give them the options, let them make educated decisions. That's what I instill in my people, Pat, and hopefully, that's what they're communicating to our clients.

PAT: Craig, the horror stories of smitten stock investors have flooded the media this last year because of hasty, presumptive investment decisions. What are some of the resources that you offer the public to help investors slow down, invest some time and do their homework before investing in U.S. rare coins... or other new market opportunities?

CRS: You bring up the question of the horror stories of smitten investors and you're so right, I mean I read an article in *Newsweek* last week that showed these various people and how they've had to cut back on trips and even sending their kids to college and stuff because they have been smitten in what's been going on in the stock market. But I really believe that what it all boils down to is education.

Many people got involved in the stock market for the wrong reason. The stock market is investment, not savings, and many people took their savings dollars and invested it in the stock market. And that's wrong. And what we do in our resources that we offer to the public, and we offer them all with no obligation. Pat, I want to educate people to show them that there is a difference between investment and savings. There's a difference between a 401K, which by the way is a great way to invest in the marketplace, but totally different than holding tangible assets.

Most people that will listen to this message that we're doing Pat, will say that the number one place that they've made money in their life is in the equity of their home. And yet most people don't even realize that that is part of your investment portfolio.

I think the key, Pat, of what we offer in our educational materials, of course we have our newsletter, and our daily updates on the web, and we have a CD, and we have a video, and we have books, as a matter of fact I'm in the process of writing a new book called Rediscovering Gold in the 21st Century. We have all those things, but what we can offer greater than that is one simple word: and that is "diversification."

You know you don't hear the talking heads on CNBC and MSNBC talk about diversifying the holdings. And if they do talk about diversification they are telling you have some tech stocks and bio-med stocks, some financial stocks. That's not a well diversified portfolio.

A diversified portfolio is to have stocks, bonds, treasury bills, cash, real estate, tangibles, and that's how you truly walk into these type of markets, get into these type of markets.

But the educational material that we really highly recommend to people, before they get involved in rare coins, is of course our Rare Coin Buyer's Guide, along with our U.S. Gold Commemorative Report, and our constantly changing, Real Money Perspectives newsletter. Our Real Money Perspectives newsletter is getting ready to celebrate its 17th year in publication. And you know every month or six weeks David Bradshaw and I really do our level best to bring a balanced approach in that newsletter of how to face these markets, and you know I'm very proud to say this.

Do you know that our philosophy is work? Look our markets haven't always been up, there've been years they've been down. And they've been down years that they were supposed to be down because when certain markets go up, certain markets go down. And now you have the stock market going down and our markets are going up. And that's the way it's supposed to be, you know, investors have to make their expectations realistic.

They became very unrealistic, I believe they almost thought that it was a birthright, if you will, in order to see double-digit returns every year, Pat. And that's just not realistic. You know, if we can get 8-12 percent growth per year on our portfolios we should be tickled to death. And that's when you have a balanced portfolio, that's exactly what you can expect.

But unfortunately, people got suckered in the early 90s, to putting all their eggs in one basket, the world-famous stock market, and although it rewarded people quite well for a number of years, it's punished people quite a bit recently.

Even great companies like Xerox, Cisco, others that have just been decimated by what's happened recently. And that doesn't mean they're not going to come back - look - everybody thought the game was over after the great crash of 1929, and the fabulous '50s, and the prosperity and growth of the '90s, clearly shot that theory down.

So we're going to have a brighter day, but there's going to be some treacherous ground that we're going to cross, getting to that brighter day. So I want the people informed, and I want the people to get our resources, get a copy of our newsletter, go to our Web site at buycoin.com, get one of our CDs, talk to one of my very, very informed brokers. These guys live, eat, sleep and breathe the markets, the geopolitical and economic makeup of the world that we live in today.

Probably an appropriate way to close this conversation is to quote a Bible scripture, from the Book of Proverbs, because I believe

Solomon was a very wise man, he said, "In a multitude of counselors, there's wisdom." And that's what we want you to do, we want
you to speak to our people, speak to other people, get our information, get a multitude of counselors, get the education, and then make
your decisions not basis your feelings, but basis education.

PAT: Thank you Craig Smith, president of Swiss America. Let's
now talk to a few of Craig's top brokers from the Swiss America
team. These men are on the front lines, helping Americans diversify
a portion of their money into tangible assets. Let's start with my
Swiss America broker Earl Brown.

Earl is the senior numismatist at Swiss America and a 20-year veter-
an of the firm next year. And, he's also a part-time educator and
classroom curriculum developer. Thanks for being with us Earl. You
speak to thousands of investors each year about protecting assets.
What is the number one question you're asked about the rare coin
market today?

EARL BROWN: Most of the folks who call us
through the near-20 years that we've been here,
have little or no understanding of what a tangible
market represents. Their question generally has to
do with whether it's real, how it works, or whether
there's value to it.

People call us out of curiosity, some of them call us because they're
worried about the economy, they're worried about safety, others are
aware that coins have been a viable investment for 50 years so
they're looking for the growth potential. More often than not the
number one question is, "How does it work?" And frankly Pat, it's
not complicated, generally I can break down how this works,
whether they work with us or not, I can break down the market,
both buying and selling, the risks and rewards, within a two-three
minute conversation with them.

PAT: Earl, let's say a person had a large amount of his money invested in Internet stocks that plummeted, what would be your advice at this point? Buy, sell or hold?

EARL: A reasonable strategy in any investing is to never buy or sell in a panic. If you follow that advice, then one would neither buy nor sell. But the follow-up to that advice is once you understand what the panic is based on, you make some educated decisions. And the educated investor right now is probably doing a bit of all three, buying, selling and holding. There are some viable stocks that have price-to-earnings ratios that are reasonable to consider long-term holds. There are others that are based on nothing but air, and this correction that we have gone through is not short-lived, some of these companies will not survive.

I would say the most significant event that appears to be on the horizon is the renewed interest in tangible assets, and it's not just our coins - it's land, it's real estate, it's things that have value that you can touch, taste, smell. We learn through the years things that don't have substance tend to over time show you that the substance didn't exist, and that's what's happening. So again, the answer to the question is you neither buy nor sell in a panic, but once you evaluate the market for what it appears to represent, there are reasons to do all three.

PAT: Can you give me an example of a client that was prepared for the recent crash in tech stocks because they diversified? How did coins protect their portfolio?

EARL: Coins in and of themselves don't protect a portfolio. What you attempt to do in putting together a diversified portfolio is to represent yourself in more than a single market. Over time, the tried and true aspect of our market is no different than people who have held land for long periods of time, or people who own real estate. It's going to have its ups and downs, but those who took advantage of the things we recommended prior to the year 2000 - to diversify and put a portion of their funds into our market benefited.

Our report will show you that Gold Commemorative coins, while the NASDAQ has dropped as much as 60, 70 and 80 percent in some issues and gone belly-up in others, have averaged 30 to 40 percent in our commemoratives the past two years, and I'm guessing we'll see quite a bit more of that growth in the years to come.

PAT: Two years ago, back in 1999 you began telling me about Gold Commemorative coins, which turned out to be the number one performer in the last two years. What do you see in the future for Gold Commemorative coins? Will a slumping economy affect them?

EARL: People look at the future at three different levels, I suppose; two of which I could make a case for coins and one of which I couldn't. It's not going to remain status quo; I think people are looking at this as to what is likely to occur to get us out of what we're in right now.

There are those who are fearful that we may go through some type of recession, and in fact there are those who talk about depression. Historically we have some precedent for that, and collectibles - not just coins - but antiques, paintings, tangibles once again, have done quite well during these times. There are always people with money - more millionaires are made during depressions than any other time.

So that's one scenario that would support putting a portion into this market, if that's what you envision. Others are talking about a scenario that involves hyper-inflation. Frankly that's when we've done our best, back in 1979-80, when we had double-digit inflation and double-digit interest rates. During the Carter administration, gold tripled and numismatics went up five- and ten-fold. There is a scenario, however, that if people envision it, that I can't make a case for our coins. If we go into the end of mankind as we know it, then you're better off with canned food and a can opener, because you can't eat your gold. I'd rather have gold than paper, even under those circumstances, but I don't have a precedent to go back to on a total collapse. So hyper-inflation - we do well. Depression, recession - historically we've done well.

PAT: In terms of timing and market cycles, is there a good, better or best time to invest?

EARL: The optimum time to invest in anything, whether it's coins or stocks, real estate, is always to buy at the bottom and sell at the top. But that's not realistic. I was always taught that if you buy within 10 percent of the bottom and sell within 10 percent of the top, you're going to make money, it doesn't matter what you're buying.

If you take a look at historically where U.S. rare coins have been, many coins are within ten percent of the bottom. Certainly the risk is not as great as the reward potential.

When you have a coin trading at $1,000 and it was at one time $10,000 - obviously coins will never go to zero as some of the paper investments do - but if it goes down to face value for a piece of gold, which is $250 or more right now, your downside is $750. But your upside, if your market is a real market - and our market has been through the years - your upside, if it just went back to where it was, is $9,000.

So when you decide which market to go into, you weigh the upside versus the downside. For those who think a downside doesn't exist in every market, they haven't taken a close look at what they're investing in. Even a CD has a downside, and it is a significant downside because you are guaranteed a loss. When you're getting 4, 5 and 6 percent on a CD, you can't make up ground on the price of goods. The cost of living is more than that, they just don't report it.

The best time to buy is when you're convinced that your upside is greater than your downside.

PAT: Thanks Earl, keep up the good work. Let's now talk with Dr. Fred Goldstein, a senior portfolio manager with Swiss America. Fred, I understand that you have been following gold bullion prices and trends for almost two decades. You've written about the "Untold story of gold" and the alleged price manipulation by central bankers, major Wall Street brokerages and others. Tell us the latest news and your perspective on the gold market in the year 2001.

DR. FRED GOLDSTEIN: When I speak to people across the country, I sense that many people are frightened. They've lost money in the recent stock market slide and they really don't know what to do. They've been told by their brokers and the pundits on network television: "Invest in equities for the long-term." Many investors expected to receive 10, 15, 20 percent a year compounded on their stocks, for the next 10 to15 years.

Now they're recognizing that was an unrealistic goal. And they're not sure what to do and are paralyzed, incapable of making certain decisions. We believe they can recoup some of the losses they took over the past two years in the equity market, by moving carefully to certain selected undervalued market areas.

I believe the gold market today is best described as a sleeping giant. Certainly statistics are showing that levels of inflation are picking up in the United States, the Consumer Price Index, the Producer Price Index, is at some of the highest levels it's been in about seven years. The prices of utilities, especially electricity, are soaring, natural gas and oil prices are significantly higher than analysts have predicted just a couple of years ago.

I think the most important factor is the government. The government is now printing money at a faster pace than they have in more than 20 years, when we had the last great inflation under the Carter and Reagan administrations.

So certainly inflation is coming back and gold is generally a barometer of inflation, and I would expect the price of gold to skyrocket. In fact, the price of gold should have skyrocketed already. There has been a recent lawsuit that the Gold Anti-Trust Action Committee

(GATA) has filed, accusing five bullion banks of manipulating and colluding, violating anti-trust laws and keeping the price of gold down. So the GATA lawsuit has gone to federal court.

There's also a major conference in May in Durban, South Africa, where members of the state department, the South African press, the South African government, the mining community will be explained how the gold price has been kept artificially low.

But we believe in any cycle, sometimes commodities will trade too low and sometimes too high. This is a perfect example of an over-sold, undervalued commodity, and with the world getting the information through the Internet about gold's suppression, we believe in the not-too-distant future, whether it be pressure from the U.S. government, the lawsuit, or political pressure from South Africa, that eventually gold will be allowed to freely trade. And when it does, many experts believe it will move above $600 an ounce. Most analysts believe from a technical standpoint and a fundamental standpoint, we've seen the bottom of the market, so I'm very encouraged about the short term for the price of gold.

PAT: Dr. Fred Goldstein has helped write a handy booklet entitled "Rare Coin Buyers Guide." Tell us Fred, what is the difference between investing in bullion and rare coins?

FRED: Well most people when they purchase bullion don't look at it as an investment. They want to hold gold to protect their paper assets, gold being a tangible item, an item that will hold its value, versus paper money, which will lose or become devalued as the government prints up more, or as inflation comes back, and we certainly know over the last 20 to 30 years that it takes more paper dollars to buy the same goods and services.

Pat, there are many experts out there who know a lot more about the gold market than I do, I was reading Frank Veneroso, who is an international consultant to major corporations as well as heads of state. He believes that three years ago the fair equilibrium price of gold was $600 an ounce. According to Frank, the manipulation of the market has been so extensive and so devious, that the first stop for the price of gold, the first leg up, is not $600 an ounce anymore, but $2,000 an ounce. That's a quote.

Recently I spoke to Bill Murphy, the President of the Gold Anti-Trust Action Committee (GATA), and he told me, he said, "Fred, the price of gold will explode, most people will not be prepared for it." He said, "Fred, you will not have enough gold coins to sell to your clients."

When people purchase rare coins, they're usually purchasing coins looking forward to appreciation, as a hedge against inflation. Many people enjoy the history, beauty, and art behind U.S. rare coins, but generally rare coins are more sensitive to inflation than the price of gold. And of late, certain rare coins have performed brilliantly even though the price of gold has languished in the area of $260 to $280 per ounce.

PAT: Let's talk about investment time frames. What if an investor is looking for asset growth in just three to five years? Should they look at rare coins, or are rare coins a longer-term investment?

FRED: I think that individuals should look at rare coins as a long-term investment. When there are aberrations in the market - as an example, the price of gold being so low right now at $260 dollars an ounce - I think looking three to five years down the road is very realistic.

It's very difficult to put a time frame on when a market will be in a "speculative" phase, or when it's very good to sell, or a strong growth phase. I'm encouraged by what I'm reading about right now in terms of the rare coin market and the price of gold. I'm looking at coins today trading at 10-, 15-, even 20-year lows. I would not be surprised, in the next year or two, to see the market really explode.

I think the gold price is a sleeping giant, it will surprise a lot of people. I don't think three to five years is too short a period of time.

PAT: We've heard a lot about the success of the U.S. Gold Commemorative coins in the last two years. Is this a fluke, or is this a foreshadowing of what's ahead for other types of U.S. rare coins? Can you name a few of your favorite coin market niches?

FRED: I do not believe the Commemorative area of the market has been a fluke at all. I think it was a very undervalued area that Swiss America was able to target two years ago. And I still believe today it's undervalued compared to the prices we've seen 10-15 years ago. I think there are other areas of the coin market that are undervalued, and I do believe it is foreshadowing what's ahead for certain areas of the coin market.

I do like proof gold and silver type coins. These areas of the market are very depressed right now, very sensitive to inflation, and I believe they'll move very nicely as silver and gold prices move higher. So I do not believe the Commemorative area of the market is going to be the only area of the rare coin market to move, I think there are really very many areas which present very good opportunities for the collector/investor.

PAT: Thank you Dr. Goldstein. Next we have Swiss America senior portfolio manager Richard Spohr. Welcome Rich, tell us, does the volatile fluctuations of the stock market have a direct effect on rare coin prices? If so, how?

RICHARD SPOHR: Over the last century, if you look at every single major period of volatility, when prices were bouncing both ways, both up and down, one of the things you'll find is that every major volatile period was followed by an increase in the price of the coin market.

In other words, coins are what we call a "lagging indicator," meaning as an example, the stock market drops a great degree, coins are going to go up the next several months. That's been the case if you look back to '73 -'74, and that was the case in '87.

In October of '87 the stock market dropped 40 percent, basically in a day and a half. Over the next 120 days the gold bullion market jumped 35 percent, moving from $375 to $500. During that same 120-day period of time, the rare coin market moved anywhere from 100 to 400 percent higher. That's why we tell investors that rare coins are an excellent market to hedge your stock portfolio during times of high market volatility.

PAT: Rich, in your article "How to Be a Millionaire" you give an example the how and why of hedging these risky financial markets? Can you explain?

RICH: One of the first things I ask new customers is if they understand what the word "hedge" means and how it applies to their investment - 99 percent of the people I speak to don't even know the definition of the word.

According to Webster's, the definition of hedge is, "to try to avoid or lessen loss by making counterbalancing investments."

In other words, a small portion; I recommend 15 percent of an investor's portfolio in stock, should be put into a hard tangible asset like the rare coin market. The idea being, what you lose in a down-market on the left hand, you're gaining back when the coins on the right hand rise in value.

A perfect hedge means if you have a market that drops and you lose 40 percent on the left hand, the right hand would move up enough to offset that. Going back to 1987, as an example, an investor who had between 15 and 20 percent of his stock portfolio in the rare coin market actually came out even, in some cases even a little bit ahead, because his coins rose enough to offset the loss he took in the stocks. I personally believe it is very important that any bet or investment that an investor makes should be hedged.

PAT: From your perspective, is it a good idea to buy gold coins during a period of deflating prices, or recession? Why?

RICH: These days on the financial talk shows, in the financial magazines and print media, there's been a lot of talk about inflation versus deflation and how investors should protect themselves in each of the different scenarios. Are we headed for recession? If so, where should my investments be? My answer to that is very simple.

Whether it's inflation, deflation or recession, you should always be hedged. There are certain financial instruments, like rare coins, which seem to do well in times of inflation. Of course we've had our best markets in times of inflation. The big run-up in late '79 was probably one of the biggest percentage gains in the shortest period of time in the history of man, when gold rallied up to $800.

During times of recession - a lot of people believe that during a recession when money is tight, people are losing jobs and there is a lot of financial turmoil, that that's a bad time to have some coins.

When you look at the numbers and check the charts, any time we've had some type of pullback, deflation, recession, all the way back to 1929, you will find one of the, if not the, top performing asset in that period of time, was the gold market, and if it's late enough in time, say from the '70s on up, you'll find that the rare coins were one of the top performing investments during that period of time.

So I guess what it all comes down to is, an investor should have his portfolio what I like to call bullet-proof. Whether we get hit with recession, inflation, depression, stock market crash, or we get a runaway stock market of all-time highs. Under any one of those scenarios, an investor should always have protection.

In other words, people should be able to go to bed at night and sleep knowing that their money is protected. Worst case scenario they may not make money, but an investor will hedge himself so that he would be assured that five years from today his net worth will remain the same as it is today.

PAT: Thanks Rich. Let's now examine a few of the tax advantages that gold and silver owners may qualify for with another Pat - Pat Mershon, head bullion trader for Swiss America's Product Sales division. Pat, tell us, how is Uncle Sam encouraging us to put away some gold and silver coins for retirement?

PAT MERSHON: For several years only gold and silver U.S. American Eagle coins were allowed as IRA investments. Since January 1, 1998, the new platinum American Eagle coin is also allowed, as well as gold, silver, platinum and palladium bullion which meets certain standards.

Through your Swiss America broker, they can walk you through the three simple steps. You establish the account, which is with a regulated trust, fund the account with regular contributions, transfers or rollovers, give the investment instructions to the trustee. Your Swiss America broker can help you with this process. You choose your own precious metals, negotiate the price, and leave the rest to the trustee.

It's really that simple, Pat. By taking this approach, Uncle Sam, because of the tax savings of your deduction, is underwriting your purchase of gold, silver, platinum and palladium. Your bullion coins will be stored at Republic National Bank of New York, the trustee will provide you with a semi-annual account statement showing all activity in your account and the fair-market value of your investments. When it's time to take distributions from your IRA, you can choose which type of precious metal investments to cash through Swiss America, or take a distribution of the actual coins.

This flexibility means you can wait for optimum market conditions before you sell.

PAT: Which type of gold or silver coins offer the best tax advantages?

PAT M: Pat, a self-directed IRA is exactly like any other IRA with one major difference. You get to choose where your IRA funds will be invested rather than just accepting what the IRA trustee or custodian offers. This gives you greater flexibility because you can choose precious metals, stocks, bonds, CDs, mutual funds, government obligations and other investments. So you don't have to limit it to gold and silver, platinum or palladium.

INVEST
IN LIBERTY

American Eagle Silver Bullion Coins

PAT: How should a person set up his or her portfolio to get the maximum advantages?

PAT M: Well Pat, I have a few clients, one in particular that has about $10,000 in gold Krugerrands from South Africa. They didn't have an IRA set up, they decided to set up a golden IRA, they decided to liquidate the gold over a five-year period $2,000 per year.

Well, they capture a tax loss each year because they bought the coins at over $300 each, so they pick up the difference between when they purchased them and when they liquidated them at the lower price, which they can use against their taxes, at the same time they're taking that money, and putting it into the IRA, and thus they get an additional tax deduction. So they were holding their gold, now they're holding their gold but Uncle Sam has underwritten their complete purchase. That's quite a sweet deal, Pat.

PAT: Thank you Pat. Now let's discuss a growing area of concern: financial privacy with Jim Berg, senior portfolio manager with Swiss America. Jim, what do your clients tell you that they like best about the financial privacy of U.S. rare coins?

JIM BURG: Pat, as you know, there is very little that we can hold or invest in that we don't have to disclose to anybody. And most of the clients that I talk to out there are very happy and excited to have an asset that they never have to report to any agency. They like the fact that there's no social security number ever required, and that it's their little secret treasure.

PAT: Let's say I'm now convinced that the best way to hold gold is in numismatic (or rare) coins. Can I exchange gold bullion for high performance rare coins? Could you walk me through the procedure?

JIM: Absolutely. It's actually very simple, Pat. All you have to do is let one of the brokers here know what you're holding - are you holding Krugerrands or Eagles, or Danish Kroners, Silver - let us know what it is, we'll come up with a value, have you ship it in, and then put together a portfolio that best suits you.

Are you looking for safety, are you looking for growth, and then we'll put something together along those lines. Then once we've put together the portfolio for you, what we'll do is guarantee those coins and prices and ship them back to you, registered, insured U.S. Mail.

PAT: Thanks Jim. I see we have Swiss America's Internet webmaster Steve Kirby. Tell us Steve, What effect is the Internet having on the rare coin market?

STEVE: I think the Internet has empowered the investor with an amazing amount of information and in researching possible investments. As it has in every industry, the Internet has brought a wealth of instantly available information to help investors choose where best to invest their monies.

Our portal Web site buycoin.com, for example, is a great place for the new investor who has no experience with rare coins to find hundreds of articles, statistical and historical archives on U.S.

coinage and the U.S. economy to help make educated decisions on investing.

PAT: I understand that you're constantly working on adding new features to the Swiss America family of web sites for public service and education. What are the site addresses and what are their different purposes?

STEVE: Well, first of all there's swissamerica.com which contains corporate information about the company such as contact information, media center and ordering information. And then there's buy-coin.com which contains research materials, broker perspective articles, as well as rotating features and weekly market updates.

And lastly there's true-wealth.com, which is centered on the document entitled True Wealth which is a unique perspective on investing, drafted by Craig Smith years ago, and still relevant today in this market. This site also contains regular updates and audio clips. And then we also have kind of an unrelated site called we-web.com, for the We-Blocker software, which is Internet filtering software to protect your family from pornography, adult content, violence, and drugs and alcohol that your children can be exposed to, even accidentally.

PAT: Thanks Steve, the Internet is such a wonderful tool. Well, this has been great! I don't know about you but I feel better equipped to face the financial challenges this year with some good old-fashioned gold in my portfolio.

Craig, after years of prosperity, the American economy is finally slowing down. Many businesses in turn are pulling back, but not Swiss America. I understand that 2001 is a year of expansion for Swiss America and you're deepening your roots in the community with a new corporate headquarters. Congratulations!

In closing Craig, tell us your dream for Swiss America during this season of change.

CRS: Pat, I don't think there's enough CD, enough audio tape, enough video tape, for me to tell you really what my dream is for Swiss America. It would take too long and probably bore a lot of people because I believe in dreams, I believe in goals, my pastor, the Reverand Tommy Barnett, he's probably the greatest dreamer of our day, and the things that he has accomplished - believing God for the unbelievable - have been nothing short of miraculous.

Let me just show you my short-term dreams, Pat, we're in the process of building a beautiful new corporate facility here in North Phoenix, it's going to be almost 15,000 square feet of super-high tech, the finest computers, the finest data lines, the finest of every-thing. When I say the finest, I mean top-drawer, Pat, because we want to be able to offer our clients first-rate service, first-rate reliability.

That's my dream, Pat, I want to increase the pleasure of the client experience, I want from the minute you call the firm, until the day you sell coins, or subsequently buy back, or whatever you do, I want it to be a top-drawer experience, Pat. But I know that that takes more than just a building and great equipment, and fine facilities, it takes people who have a heart, who want to help other people. And my dream is to have all my staff, five days a week, eight hours a day, focus their attention 100 percent on the client.

I'm convinced that corporations, great corporations that are going to withstand the test of time - you know Pat we're getting ready to celebrate our 20th anniversary soon - I believe that they're built on great client relationships. That we care about our clients, that we do our level best for our clients. I said this to you earlier in our conver-sation, Pat - we can't control the markets. We use our best data that's available. Look, any person that believes a stock broker can control the market or a 401K manager can control the market, or a coin broker, they're fooling themselves, Pat. We can't. But what we can do is take the best data that we can find, process it properly and come up with certain conclusions that over the long-term will play out. We're already starting to see the course that we set clear back in the early '90s is starting to come to pass now. And we're grateful for that. But what we can control is the client experience. We can control how the people at Swiss America respond to the needs of our clients. And that's what we're committed to, Pat.

My dream is the 30-some-odd thousand clients that we've dealt with throughout the years will be satisfied clients, will be clients that had a wonderful experience. You know, we've come to find out that our best source of new clients is our existing clients. People love telling people about their experiences at Swiss America. My dream is far beyond just our building and hiring more staff, and servicing more of the 270 million Americans in the country. It's to create a company that leaves a legacy that says, "That company, it's owner and its people, they really cared about people." I believe that if I can leave that legacy Pat, then I've succeeded in life. That's what my dream is.

7.2 Timeless Values - My Clients Speak Out

What do Americans want most from their investments? Many would agree that the most important elements of investing are: safety, quality, reliability and loyalty.

In the investment world, gold offers these same qualities. So why is the public hesitant in buying gold? I suspect that market values are changing so quickly today that many have lost their moorings.

For nearly twenty years, I have advised clients to diversify a small portion of their assets into U.S. gold coins for three basic reasons: safety, privacy and profit - in that order. I have reminded clients that gold represents timeless value.

Tens of thousands of our clients have taken our advice over the years and I wanted you to hear what they have to say - because they are the reason we are here. If you are looking for safety, quality, reliability and loyalty, then read on to discover why Swiss America is outstanding in our field.

I. Safety
"Numismatics: a Store of Value"

"I felt that the state of the world economy is so uncertain now, that I needed numismatics as a stable store of value in my portfolio. And I contacted several firms before selecting Swiss America and have certainly not regretted my choice. I've dealt with my broker at Swiss America several years now and their honesty and pledge of service has never been doubted. I'm extremely pleased. I've mentioned it to several friends and others and recommended the firm because of their honesty and trust." -Peg T.

"I Did Need To Diversify"

"I began investing with Swiss America a number of years ago mainly because of the direction our economy was going. It was hard to get a read on where you should be investing your assets. One thing I was quite certain of, I did need to diversify - so one direction we did go, because of the ups and downs of the economy - I made the choice to invest in hard assets due to the confidence that I had in Swiss America." -Tom Buchanan

"Leery of the Stock Market"

"We were both leery of the stock market. It seemed that if we made money, it was kind of by accident. I couldn't see any logic to it. We watched it for a while and then just decided to get out. Then what? What do you do? So, we put it in coins and at least it will hold its value. It was something to try to protect the money that I scratched together over the years. What gets me about the gold coins is how beautiful they are… how pretty they are. There seems to be somewhat of an attraction toward gold, like there is sitting around looking at a campfire - like they take us back to our roots or our ancestors or something. They are encapsulated so they can't get scuffed or roughed up." -Mr. C.

II. Quality
"U.S. Coins: No Longer a Mystery"

"I considered the coin market pretty much a mystery until I had investigated it with you people. I had bought a book from a conservative publisher and specifically, Swiss America was mentioned as being a very high quality firm with high ethical standards and committed to an honest presentation of the collectibles market. In an area where gold has been surrounded by myth and greed, it's hard to find out who's really an authentic dealer. It was an area where I had no expe-

rience, so it provoked a phone call and then it became clear to me that without some tangible assets, playing the market, the stock market or the bond market, was riskier.

"So, I wanted to reduce my risk and have a tangible base in there somewhere. You clearly described the superiority of the collectibles as having their rarity as a prime factor in terms of their valuation. The bullion market would seem to me to be closer to a gold price function, whereas, the rarity value of the collectibles ends up being the most important factor. And as we progress over the years, that rarity value becomes even more important. It's like a built in level of strength that makes the collectibles much more desirable. I had put in an initial amount of somewhere around $25,000 and it has grown - almost doubled, I guess in the last two or two-and-a-half years. I'm quite satisfied with it." -Jon Z.

"The Professionalism Sold Me"
"I realized the need for the security, particularly facing the changing times that we're in. I feel very, very good about the information that has been furnished by the brokers and the other people. The professionalism and the expert way that they furnished me the information, probably is what really sold me on it. I feel like I was led into a really good thing." -Mary K.

III. Reliability
"Coin Performance as Expected"
"I chose collectible gold because of the safety factors. I felt that it would be something that I could use in case of a monetary breakdown - that I would actually have in my hand. I am very satisfied with my purchases and my coins have done pretty much exactly what Swiss America has told me that they would do. Their sales, service and delivery have been excellent. I've referred several of my close friends because I would like to see them participate in what I feel to be an excellent investment." -Darvin F.

"A Wise Purchase"
"I just heard their number on the radio. I gave them a call, and gave some other ones a call and after I talked to everybody, I compared my notes and I was most happy with Swiss America. The broker I had, I was real pleased with what he had to say and how he handled everything. I got immediate service; everything was delivered on time. I've been doing business with you for some time now and

I've always liked the service I've gotten and I like the quality of the product. They're almost matching the stock market across the board, plus I've got a solid asset to them that you don't have in paper. They are a wise purchase." -Mr. H

IV. Loyalty
"A Legacy For My Kids... Now or Later"
"The representatives at Swiss America are A-1. They've always been free with the information. My coins arrived on time, just like they said they would. They all came in A-1 condition and I've been real proud of every one I have and I still have every one of them. I have not sold any of them and don't intend to. These coins are basically going to be a legacy for my kids - something I can give them before, or they can get it after." -Mrs. A.

"My Only Regret"
"To anyone else that is considering investing in gold, I would recommend them to you. Swiss America is a company that, over the last number of years, has proved to be very reputable. The only thing I regret is that I don't have more money to invest in these beautiful gold coins. I'm really excited!" -Mr. B.

7.3 A Final Word
"Thou Shalt Diversify"

They say a man's last words in life (or in a book) are the most important - summing up the knowledge, wisdom and experience learned in life. So here goes.

"Thou shalt diversify," is my number one commandment to investors. I have lived by this message for more than two decades and it has saved my skin - and portfolio - again and again. By spreading your money into a diversity of asset types you are escaping today's stressful investment roller coaster.

Diversification is also the best way to become a charter member of the "No Whiners Club," an elite group of investors who have stopped blaming their brokers, the market, the government or Mr. Greenspan for sudden unexpected market losses due to wild speculation and volatility.

Yes, the next gold rush has already started, though unannounced in the mass media. But very soon the news will reach Main Street that even the supposedly outdated market for gold bullion has outperformed the NASDAQ with a modest 8 percent rise between June 1999 and June 2001. Can you imagine what will happen to rare gold coins during the next global gold rush?

I can. I watched in awe during the last gold rush ('79-'80) and trust me when I say that the time to acquire gold and silver is now, not later. Gold is one of the only assets that is not simultaneously someone else's liability, and the world's economic leaders know it. For that reason alone everyone should own some gold.

"Owe no man anything..." is another truism to help achieve financial liberty. Perhaps you are laboring under a heavy burden of debt today which prevents you from investing in anything. My best advice is to develop a plan to rid yourself of all non-productive debt as quickly as possible. Start saving something every month for the day that you can no longer save.

The Privacy Factor

We Americans have a tendency to believe that we have nothing to hide. If a policeman knocks on your door or wants you to open the trunk of your car, you say, "Welcome, I have nothing to hide." Unfortunately, we've unknowingly taken respect for authority into the financial world.

Americans enjoy unlimited benefits from new technologies in a wired world. But those wires send information in two directions, and the access to our personal data has never been more open for abuse. It's not just the Internet that erodes our privacy.

In dozens, possibly hundreds, of every-day activities, you leave a trail of who you are. As technology brings us closer together, the fragments of information about you are becoming much easier to piece together, revealing the most intimate details of your life.

The crowning advantage of owning U.S. rare gold coins is that they are 100 percent private. There is no registering of U.S. rare coins. Stocks, bonds, mutual funds and treasury bills are all registered investments.

In the case of a stock, if you buy it for $10 a share and then you sell it for $20 (or $5) a share, the stock brokerage must report that gain to the federal government. Not so in the case of rare (or numismatic) gold coins - whether you buy or sell, there is no transaction report filed with the state or federal government.

Our Founding Fathers invested their highest collective wisdom into our precious founding documents to ensure America would never be taken captive by tyrannical forces - governmental or economic - as long as we obey the law.

For that reason I've enclosed our original founding documents in the Appendix. I suggest taking an hour and reading carefully the law of the land. As they say, "Ignorance of the law is no defense." Political, social and economic liberty are the bedrock of America's unique brand of freedom - just like gold is the bedrock of America's unique history of economic freedom.

The bottom line is that historic United States gold and silver coins stand as one of the last guardians of your financial freedom, liberty and privacy in the 21st century - just as they've done in the 18th, 19th and 20th century.

So, let's celebrate freedom by rediscovering gold. I hope this book has helped you to better understand why it is so critical to have a golden anchor in your portfolio. Now you can have fun, make money and discover America's rich heritage - all at the same time.

APPENDIX

I. Historic Quotables

A collection of quotes from America's Founding Fathers, selected economists and a few notable contemporary leaders.

AMERICA'S FOUNDING FATHERS

All the perplexities confusion and distress in America arise not from defects of the Constitution, not from want of honor or virtue, so much as from downright ignorance of the nature of coin, credit and circulation. **-John Adams,** in a letter to Thomas Jefferson in 1787

If the American people ever allow private banks to control the issue of their currency, first by inflation and then by deflation, the banks and corporations that will grow up around them will deprive the people of all property until their children wake up homeless on the continent their fathers conquered.
-Thomas Jefferson

Of all the contrivances devised for cheating the laboring classes of mankind, none has been more effective than that which deludes him with paper money.
-Daniel Webster

The colonies would have gladly born the little tax on tea, and other matters, had it not been that England took away from the colonies their money.
-Benjamin Franklin

This is a favorable moment to shut and bar the door against paper money. The mischief of the various experiments which have been made are now fresh in the public mind and have excited the disgust of all the respectable parts of America.
-Oliver Ellsworth, a delegate from Connecticut, who later became this nation's third Chief Justice of the Supreme Court.

I believe that banking institutions are more dangerous to our liberties than standing armies. Already they have raised up a money aristocracy that has set the government at defiance. **-Thomas Jefferson**, at the Constitutional Convention (1787)

It's apparent from the whole context of the Constitution as well as the history of the times which gave birth to it, that it was the purpose of the Convention to establish a currency consisting of the precious metals. These were adopted by a permanent rule excluding the use of a perishable medium of exchange, such as of certain agricultural commodities recognized by the statutes of some States as tender for debts, or the still more pernicious expedient of paper currency.
-President Andrew Jackson, 8th Annual Message to Congress (December 5, 1836)

If what is used as a Medium of exchange is fluctuating in its Value it is no better than unjust Weights and measures, both which are condemned by the laws of GOD and Man, and therefore the longest and most universal Custom could never make the Use of such a Medium either lawful or reasonable.
-Roger Sherman, a delegate from Connecticut and author of the gold and silver coin provision of the Constitution, wrote a scathing condemnation of paper money entitled "A Caveat (caveat means warning) Against Injustice"

CURRENCY/CREDIT

What is robbing a bank compared with founding a bank? **-Bertolt Brecht,** The Threepenny Opera

Man can live and satisfy his wants only by ceaseless labor; by the ceaseless application of his faculties to natural resources. This process is the origin of property. But it is also true that a man may live and satisfy his wants by seizing and consuming the products of the labor of others. This process is the origin of plunder. Now since man is naturally inclined to avoid pain - and since labor is pain in itself - it follows that men will resort to plunder whenever plunder is easier than work. When plunder becomes a way of life for a group of men living in society, they create for themselves, in the course of time, a legal system that authorizes it and a moral code that glorifies it. **-Frederick Bastiat**, Economist, Statesman,

Give me control over a nation's currency and I care not who makes its laws.
-Baron M.A. Rothschild

Whoever controls the money in any country is master of all its legislation and commerce. **-President James Garfield**

Centralization of credit in the hands of the state, by means of a national bank with state capital and an exclusive monopoly. **-Karl Marx**, 5th Plank of the Communist Manifesto(1848)

We have in this country one of the most corrupt institutions the world has ever known. I refer to the Federal Reserve Board and the Federal Reserve Banks. Some people think the Federal Reserve Banks are U.S. government institutions. They are not government institutions. They are private credit monopolies; domestic swindlers, rich and predatory money lenders which prey up on the people of the United States for the benefit of themselves and their foreign customers. The Federal Reserve Banks are the agents of the foreign central banks. The truth is the Federal Reserve Board has usurped the Government of the United States by the arrogant credit monopoly which operates the Federal Reserve Board.
-75th Congressional Record 12595-12603

It is well enough that the people of the nation do not understand our banking and monetary system, for if they did, I believe there would be a revolution before tomorrow morning. **-Henry Ford**

MONEY

Silver and gold are not the only coin; virtue too passes current all over the world.
-Edipus

Inflation is the one form of taxation that can be imposed without legislation.
-Milton Friedman, Nobel Prize winning economist

Never ask of money spent - Where the spender thinks it went. Nobody was ever meant - To remember or invent -What he did with every cent.
-Robert Frost

The chief value of money lies in the fact that one lives in a world in which it is overestimated. **-H. L. Mencken,** Writer, editor, social critic

Honesty is the best policy, when there is money in it. **-Mark Twain**

Money often costs too much. **-Ralph Waldo Emerson**

Today you can go to a gas station and find the cash register open and the toilets locked. They must think toilet paper is worth more than money.
-Joey Bishop, Comedian

Money is like a sixth sense, and you can't make use of the other five without it.
-W. Somerset Maugham, Playwright, Of Human Bondage (1915)

Money does all things for reward. Some are pious and honest as long as they thrive upon it, but if the devil himself gives better wages, they soon change their party. **-Seneca,** (4B.C.-65A.D.) Playwright, Orator, Philosopher

He that is of the opinion money will do everything may well be suspected of doing everything for money. **-Benjamin Franklin**

Put not your trust in money, but put your money in trust.
-Oliver Wendell Holmes, (1809-1894) American writer, teacher

People who never do more than they get paid for, never get paid for anything more than they do. **-Kemmons Wilson,** Founder, Holiday Inns

Those who have some means think that the most important thing in life is love, the poor know that it is money. **-Gerald Brenan,** English writer

A frustrated manager complained that every time he provided training, the now highly skilled employee was snatched up by a competitor: The only thing worse than training people and having them leave is not training them and having them stay. **-Zig Ziglar,** Author, Orator

If you can actually count your money you are not really a rich man.
-J. Paul Getty, Oil Tycoon

Pennies do not come from heaven, they need to be earned here on earth.
-Margaret Thatcher

Money. You can be young without it, but you can't be old without it.
-Tennessee Williams

Gold, like the sun, which melts wax and hardens clay, expands great souls and contracts bad hearts. **-Antoine de Rivaroli** (1753-1801)

Our system of credit is concentrated. The growth of the nation, therefore, and all our activities are in the hands of a few men. We have come to be one of the ruled, one of the most completely controlled and dominated governments in the civilized world - no longer a government by free opinion, no longer a government by conviction and the vote of the majority, but a government by the opinion and duress of small groups of dominant men. **-Woodrow Wilson,** three years after signing the Federal Reserve Act into law.

Money is a good servant but a poor master. **-Dominique Bouhours,** Author, French Jesuit (1632-1702)

If money could talk, it would say Good-bye. **-Unknown**

The Treasury prints "In God We Trust" on our currency. God is indeed trustworthy, but trusting an intrinsically worthless currency takes an awful lot more than faith. **-Peter Kershaw,** Economic Solutions

A Federal Reserve Note [is] merely an IOU. Here's how it works. When the politicians want more money, they dispatch a request to the Federal Reserve for whatever sum they desire. The Bureau of Printing and Engraving then prints up bonds indenturing taxpayers to redeem their debts. The bonds are then 'sold' to the Federal Reserve. But not this unusual twist-the bonds are paid for with a check backed by nothing! It is just as if you were to look into your account and see a balance of $412 and then, hearing that government bonds were for sale, write a draft for $1 billion. Of course, if you did that, you would go to jail. The bankers do not. In effect, they print the money that enables their check to clear. **-James Dale Davidson**, Director, National Taxpayer's Union

The one aim of these financiers is world control by the creation of inextinguishable debts. **-Henry Ford**

The privilege of creating and issuing money is not only the supreme prerogative of Government, but is the Government's greatest creative opportunity. By the adoption of these principles, the taxpayers will be saved immense sums of interest. **-President Abraham Lincoln**

About all a Federal Reserve note can legally do is wipe out one debt and replace it with itself, another debt; a note that promises nothing. If anything has been paid, the payment occurs only in the minds of the parties-in the ideaspere-not the real world. **-Tupper Saucy**, The Miracle on Main Street

A great fortune is a great slavery. **-Seneca**

Inflation is a method of taxation which the government uses to secure the command over real resources; resources just as real as those obtained by ordinary taxation. What is raised by printing notes is just as much taken from the public, as is an income tax. A government can live by this means, when it can live by no other. It is the form of taxation that the public finds hardest to evade, and even the weakest government can enforce it when it can enforce no other. By a continuous process of inflation, government can confiscate secretly and unobserved an important part of the wealth of their citizens. By this method, they not only confiscate, they confiscate arbitrarily, and while the confiscation impoverishes many, it enriches some. Lenin was certainly right, 'there is no surer way of overturning a society, than to debauch the currency.' The process engages all the hidden forces of economic law on the side of destruction, and does o in such a manner than only one man in a million is able to diagnose it.
-John Maynard Keynes, The Economic Consequences of Peace (credited for the birth of modern economics).

The Bureau prints approximately sixteen million notes each day. The Bureau has the power to create money and almost any amount of it. The only limiting factors are the speed of the presses, and the public's willingness to accept it.
-Government Securities, U.S. Bureau of Engraving

Without the confidence factor, many believe a paper money system will eventually collapse. Present experience indicates the system can operate without a gold guarantee however, and that they only confidence required is a firm conviction that money will be accepted in payment for goods and services. **-Gold,** Federal Reserve Bank of Philadelphia

Government is the only agency that can take a valuable commodity like paper, slap some ink on it, and make it totally worthless. **-Ludwig von Mises,** Austrian Economist

A good man leaves an inheritance for his children's children. **-Proverbs 13:22**

Strictly speaking, the 'government' of the United States (or of any State or locality) is a kind of 'legal fiction.' It is not the individuals elected or appointed to office, the physical buildings they occupy, or the actions they take per se. Rather, the 'government,' rightly understood, is the set of actions duly elected or appointed officials take that are consistent with the Constitution. If an action is inconsistent with the Constitution, it is unlawful and non-governmental, by definition. Such as unconstitutional action may be defined as usurpation or tyranny, but never as a truly governmental act. Simply put, our government has no authority to act outside of or against the Constitution; and when public officials do so, they are not acting as agents of government, but as lawbreakers or outlaws. [I]n the most fundamental sense, the United States need no 'reform' law, or 'restoration' law, to return to sound money. For the necessary law already exists, in the Constitution itself. What stands in the way of monetary freedom...is not law, but lawlessness; not government, but usurpation and tyranny. **-Dr. Edwin Vieira, Jr.,** Director, National Alliance for Constitutional Money

In the beginning of a change, the patriot is a scarce man; brave, hated, and scorned. When his cause succeeds, however, the timid join him, for then it costs nothing to be a patriot. **-Mark Twain**

FRUGALITY

A miser grows rich by seeming poor; an extravagant man grows poor by seeming rich. **-William Shenstone,** (1714-1763) Author, Poet

Whatever you have, spend less. -**Samuel Johnson,** 18th Century English Author (most quoted next to Shakespeare).

I was part of that strange race of people aptly described as spending their lives doing things they detest to make money they don't want to buy things they don't need to impress people they dislike. -**Emile Henry Gauvreay**

A small debt produces a debtor; a large one, an enemy. -**Publius Syrus** (42 B.C.)

People come to poverty in two ways: accumulating debts and paying them off. -**Jewish Proverb**

Annual income twenty pounds, annual expenditure nineteen nineteen and six, result happiness. Annual income twenty pounds, annual expenditure twenty pounds ought and six, result misery. -**Charles Dickens**, David Copperfield

Every man is the architect of his own fortune. -**Appius Claudius Caecus**, 'Speech to Caesar on the State'

Money was made, not to command our will, But all our lawful pleasures to fulfill. Shame and woe to us, if we our wealth obey; The horse doth with the horseman away. -**Abraham Cowley,** 17th Century English Writer

If you'd know the value of money, go and borrow some. -**Benjamin Franklin**, Poor Richard's Almanac

A bank is a place that will lend you money if you can prove that you don't need it. -**Bob Hope**, Life in the Crystal Palace

Money is a handmaiden, if thou knowest how to use it; a mistress, if thou knowest not. -**Horace,** (65 B.C. -8 B.C.) Poet, Roman Philosopher

No man's credit is as good as his money. -**Edgar Watson Howe** (1853-1937)

What makes all doctrines plain and clear? About two hundred pounds a year. And that which was proved true before, prove false again? Two hundred more. -**Samuel Johnson**

When it is a question of money, everybody is of the same religion. **-Voltaire**

One man's wage increase is another man's price increase. **-Harold Wilson**
(1916-1995) British Statesman

Labor is prior to, and independent of, capital. Capital is only the fruit of labor, and could never have existed if labor had not first existed. Labor is the superior of capital, and deserves much the higher consideration. **-Abraham Lincoln**

When a man tells you that he got rich through hard work, ask him: 'Whose?'
-Don Marquis

Giving money and power to government is like giving whiskey and car keys to teenage boys. **-P.J. O'Rourke**

A feast is made for laughter, and wine maketh merry: but money answereth all things. **-Ecclesiastes 10:19**

The love of money is the root of all evil. **-I Timothy 6:10**

Human relationships transcend currencies. **-Dennis Peacocke,** Strategist

A man is usually more careful of his money than he is of his principles.
-Edgar Watson Howe

Money is a singular thing. It ranks with love as man's greatest source of joy. And with death as his greatest source of anxiety. **-John Kenneth Galbraith,** The Age of Uncertainty

Money couldn't buy friends but you got a better class of enemy.
-Spike Milligan, Puckoon

Today we have an occult money system, no doubt about it.
-R.E. McMaster Jr., Economist, The Reaper

Never invest your money in anything that eats or needs repainting. **-Billy Rose,** in New York Post

There was a time when a fool and his money were soon parted, but now it happens to everybody. **-Adlai E. Stevenson**, The Stevenson Wit

There are two times in a man's life when he should not speculate: when he can't afford it, and when he can. **-Mark Twain**, Following the Equator

ECONOMICS

A study of economics usually reveals that the best time to buy anything is last year. **-Marty Allen**

It's a recession when your neighbor loses his job; it's a depression when you lose your own. **-Harry S. Truman**

The individual serves the industrial system not by supplying it with savings and the resulting capital; he serves it by consuming its products. **-John Kenneth Galbraith**, The New Industrial State

Inflation is like sin; every government denounces it and every government practices it. **-Frederick Leith-Ross**, The Observer

An economist is a man who states the obvious in terms of the incomprehensible. **-Alfred A. Knopf**

Debt is the fatal disease of all republics, the first thing and the mightiest to undermine governments and corrupt the people. **-Phillips, Wendell**

The only function of economic forecasting is to make astrology look respectable. **-John Kenneth Galbraith**

When the President signs this act, the invisible government by the money power will be legalized. **-Con. Charles A. Lindbergh, Sr.**, father of the famed aviator

At one time bankers were merely middlemen. They made a profit by accepting gold and coins brought to them for safekeeping and lending them to borrowers.

But they soon found that the receipts they issued to depositors were being used as a means of payment. These receipts were acceptable as money since whoever held them could go to the banker and exchange them for metallic money.
-Modern Money Mechanics, published by the Federal Reserve Bank of Chicago, page 4.

The definition of the word "dollar" has undergone such a transformation to hide the fact that it is not money, but a unit of measurement for gold and silver coin. Logically, if there are no gold and silver coins, there are no dollars of anything. Dollars cannot be money any more than quarts can be milk. A unit of measurement cannot replace or become the "thing" for which it is a measure.

However, in the mind of the public, this is exactly what has happened. People have been led to believe that a dollar is both money and a measure of it. This is what George Orwell called "double-think," where the mind is infiltrated with conflicting concepts. A dollar unit of paper money that is not one hundred percent redeemable in gold or silver coin is a dollar unit of inflation, which is a dollar unit of credit, which is a dollar unit of NOTHING.

The sole function of paper money that is not one-hundred percent redeemable in gold or silver coin is to get things without paying for them. Those who issue and control paper money as credit get everything for nothing. How the U.S. Government borrows money from the Federal Reserve is not much different from the Abbott and Costello comedy sketch in their movie "Buck Privates" where Bud Abbott wants to borrow fifty dollars from Lou Costello.

Bud:	*Do me a favor. Loan me fifty dollars.*
Lou:	*I can't lend you fifty dollars.*
Bud:	*Yes, you can.*
Lou:	*No, I can't. All I got is forty dollars.*
Bud:	*All right. Give me the forty dollars and you owe me ten.*
Lou:	*O.K. I owe you ten.*
Bud:	*That's right.*
Lou:	*How come I owe you ten?*
Bud:	*What did I ask you for?*

Lou: Fifty.

Bud: And how much did you give me?

Lou: Forty.

Bud: So you owe me ten dollars.

Lou: That's right. Well, you owe me forty.

Bud: Now, don't change the subject.

Lou: I'm not changing the subject. You're trying to change my finances. Come on now. Give me my forty dollars.

Bud: All right. There's your forty dollars. Give me the ten dollars you owe me.

Lou: I'm paying you on account.

Bud: On account?

Lou: On account I don't know how I owe it to you.

Bud: That's the way you feel about it. It's the last time I'll ever ask you for the loan of fifty dollars.

Lou: Wait a minute, Smitty. How can I loan you fifty dollars now? All I have is thirty.

Bud: Well, give me the thirty, and you owe me twenty.

Lou: O.K., this is getting worse all the time. First, I owe him ten; now I owe him twenty.

Bud: Well, why do you run yourself into debt?

Lou: I'm not running in. You're pushing me.

Bud: I can't help it if you can't handle your finances. I do all right with my money.

Lou: And you're doing all right with mine too.

Bud: Now, wait a minute. I ask you for the loan of fifty dollars and you gave me thirty. So you owe me twenty dollars. Twenty and thirty is fifty.

Lou: No, no, no...

-Steven Jacobson, Journalist, Wake Up America

II. OUR FOUNDING DOCUMENTS

1. The Declaration of Independence

2. The Constitution of the United States

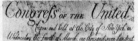

3. The Bill of Rights

GENERAL INDEX:

THE UNANIMOUS DECLARATION

I.   Need to dissolve certain political relationships.
II. Need to assume powers which God entitles man.
III. Declaring separation from unjust government.
IV. Self-evident truths elaborated.
    A. All men are created equal.
    B. God our Creator gives to each unalienable Rights
      1. Life, Liberty, Happiness, property, safety, respect, privacy, etc.
    C. The purpose of government is to protect the weak from the strong.
    D. Right and duty to abolish bad government.
      1. Fact: The Revolution was not out of rebellion by the colonies, but rather England rebelled against God's Law by repeated injuries of usurpation and tyranny. The young colonies were forced to defend themselves against the King's tyranny.
        a. eg. Bad laws, bad courts, police state (swarms of soldiers), taxes without consent, deprived of trial by jury, deporting people for trial. England declared the colonies out of their protection, rights of individuals plundered.
        b. The colonies repeatedly petitioned England, but only received repeated injury.
        c. England was warned from time to time.
        d. England was deaf to the voice of justice.
V.   The colonies appealed to God, the Supreme Judge of the world.
VI. The colonies right to be free and independent.
VII. Under the protection of God they pledged their lives, fortunes and honor.

Complete Document:
The Declaration of Independence
[Adopted in Congress 4 July 1776]

The Unanimous Declaration of the Thirteen United States of America

When, in the course of human events, it becomes necessary for one people to dissolve the political bonds which have connected them with another, and to assume among the powers of the earth, the separate and equal station to which the laws of nature and of nature's God entitle them, a decent respect to the opinions of mankind requires that they should declare the causes which impel them to the separation.

We hold these truths to be self-evident, that all men are created equal, that they are endowed by their Creator with certain unalienable rights, that among these are life, liberty and the pursuit of happiness. That to secure these rights, governments are instituted among men, deriving their just powers from the consent of the governed. That whenever any form of government becomes destructive to these ends, it is the right of the people to alter or to abolish it, and to institute new government, laying its foundation on such principles and organizing its powers in such form, as to them shall seem most likely to effect their safety and happiness. Prudence, indeed, will dictate that governments long established should not be changed for light and transient causes; and accordingly all experience hath shown that mankind are more disposed to suffer, while evils are sufferable, than to right themselves by abolishing the forms to which they are accustomed.

But when a long train of abuses and usurpations, pursuing invariably the same object evinces a design to reduce them under absolute despotism, it is their right, it is their duty, to throw off such government, and to provide new guards for their future security. -- Such has been the patient sufferance of these colonies; and such is now the necessity which constrains them to alter their former systems of government. The history of the present King of Great Britain is a history of repeated injuries and usurpations, all having in direct object the establishment of an absolute tyranny over these states. To prove this, let facts be submitted to a candid world.

He has refused his assent to laws, the most wholesome and necessary for the public good.

He has forbidden his governors to pass laws of immediate and pressing importance, unless suspended in their operation till his assent should be obtained; and when so suspended, he has utterly neglected to attend to them.

He has refused to pass other laws for the accommodation of large districts of people, unless those people would relinquish the right of representation in the legislature, a right inestimable to them and formidable to tyrants only.

He has called together legislative bodies at places unusual, uncomfortable, and distant from the depository of their public records, for the sole purpose of fatiguing them into compliance with his measures.

He has dissolved representative houses repeatedly, for

opposing with manly firmness his invasions on the rights of the people.

He has refused for a long time, after such dissolutions, to cause others to be elected; whereby the legislative powers, incapable of annihilation, have returned to the people at large for their exercise; the state remaining in the meantime exposed to all the dangers of invasion from without, and convulsions within.

He has endeavored to prevent the population of these states; for that purpose obstructing the laws for naturalization of foreigners; refusing to pass others to encourage their migration hither, and raising the conditions of new appropriations of lands.

He has obstructed the administration of justice, by refusing his assent to laws for establishing judiciary powers.

He has made judges dependent on his will alone, for the tenure of their offices, and the amount and payment of their salaries.

He has erected a multitude of new offices, and sent hither swarms of officers to harass our people, and eat out their substance.

He has kept among us, in times of peace, standing armies without the consent of our legislature.

He has affected to render the military independent of and superior to civil power.

He has combined with others to subject us to a jurisdiction foreign to our constitution, and unacknowledged by our laws; giving his assent to their acts of pretended legislation:

For quartering large bodies of armed troops among us:

For protecting them, by mock trial, from punishment for any murders which they should commit on the inhabitants of these states:

For cutting off our trade with all parts of the world:

For imposing taxes on us without our consent:
For depriving us in many cases, of the benefits of trial by jury:

For transporting us beyond seas to be tried for pretended offenses:

For abolishing the free system of English laws in a neighboring province, establishing therein an arbitrary government, and enlarging its boundaries so as to render it at once an example and fit instrument for introducing the same absolute rule in these colonies:

For taking away our charters, abolishing our most valuable laws, and altering fundamentally the forms of our governments:

For suspending our own legislatures, and declaring themselves invested with power to legislate for us in all cases whatsoever.

He has abdicated government here, by declaring us out of his protection and waging war against us.

He has plundered our seas, ravaged our coasts, burned our towns, and destroyed the lives of our people.

He is at this time transporting large armies of foreign

mercenaries to complete the works of death, desolation and tyranny, already begun with circumstances of cruelty and perfidy scarcely paralleled in the most barbarous ages, and totally unworth the head of a civilized nation.

He has constrained our fellow citizens taken captive on the high seas to bear arms against their country, to become the executioners of their friends and brethren, or to fall themselves by their hands.

He has excited domestic insurrections amongst us, and has endeavored to bring on the inhabitants of our frontiers, the merciless Indian savages, whose known rule of warfare, is undistinguished destruction of all ages, sexes and conditions.

In every stage of these oppressions we have petitioned for redress in the most humble terms: our repeated petitions have been answered only by repeated injury. A prince, whose character is thus marked by every act which may define a tyrant, is unfit to be the ruler of a free people.

Nor have we been wanting in attention to our British brethren. We have warned them from time to time of attempts by their legislature to extend an unwarrantable jurisdiction over us. We have reminded them of the circumstances of our emigration and settlement here. We have appealed to their native justice and magnanimity, and we have conjured them by the ties of our common kindred to disavow these usurpations, which, would inevitably interrupt our connections and correspondence. We must, therefore, acquiesce in the necessity, which denounces our separation, and hold them, as we hold the rest of mankind, enemies in war, in peace friends.

We, therefore, the representatives of the United States of America, in General Congress, assembled, appealing to the Supreme Judge of the world for the rectitude of our intentions, do, in the name, and by the authority of the good people of these colonies, solemnly publish and declare, that these united colonies are, and of right ought to be free and independent states; that they are absolved from all allegiance to the British Crown, and that all political connection between them and the state of Great Britain, is and ought to be totally dissolved; and that as free and independent states, they have full power to levy war, conclude peace, contract alliances, establish commerce, and to do all other acts and things which independent states may of right do. And for the support of this declaration, with a firm reliance on the protection of Divine Providence, we mutually pledge to each other our lives, our fortunes and our sacred honor.

GENERAL INDEX TO:
THE CONSTITUTION OF THE UNITED STATES

Preamble: The people hold the power, "We the people...in order to form a more perfect union...and secure the blessings of liberty..."

ARTICLE I
SECTION:
1. Legislative powers.
2. House of representatives; qualification of members; apportionment of representatives and direct taxes; census; first apportionment; vacancies; officers of the house; impeachments.
3. Senate: classification of senators; qualifications of; vice president to preside; other officers; trial of impeachments.
4. Election of members of congress; assembling of congress.
5. Powers of each house; punishment for disorderly Behaviour; journal; adjournments.
6. Compensation and privileges; disabilities of members.
7. Revenue bills; passage/approval of bills; orders/resolutions.
8. General powers of congress; borrowing of money; regulations of commerce; naturalization and bankruptcy; money; weights and measures; counterfeiting; post offices; patents and copyrights; inferior courts; piracies and felonies; war; marque and reprisal; armies; navy; land and naval forces; calling the militia; District of Columbia; to enact laws necessary to enforce the Constitution.
9. Limitations of congress; imigration; writ of habeas corpus; bills of attainder and ex post facto laws prohibited; direct axes; exports not to be taxed; interstate shipping; drawing money from the treasury; financial statements to be published; titles of nobility and favors from foreign powers prohibited.
10. Limitations of the individual states; no treaties; letters of marque and; no coining of money; bills of credit; not allowed to make any Thing but gold and silver Coin a tender in payment of debts; no bills of attainder; ex post facto Law or law impairing the obligation of contracts; no titles of nobility; state imposts and duties; further restrictions on state powers.

Complete Document:

Constitution

of the

United States of America

We the People of the United States, in Order to form a more per-fect Union, establish Justice, insure domestic Tranquility, provide for the common defense, promote the general Welfare, and secure the Blessings of Liberty to ourselves and our Posterity, do ordain and establish this Constitution for the United States of America.

Article. I.

Section. 1. All legislative Powers herein granted shall be vested in a Congress of the United States, which shall consist of a Senate and House of Representatives.

Section. 2. The House of Representatives shall be composed of Member chosen every second Year by the People of the several States, and the Electors in each State shall have the Qualifications requisite for Electors of the most numerous Branch of the State Legislature.

No Person shall be a Representative who shall not have attained to the Age of twenty five Years, and been seven Years a Citizen of the United States, and who shall not, when elected, be an Inhabitant of that State in which he shall be chosen.

Representatives and direct Taxes shall be apportioned among the several States which may be included within this Union, according to their respective Numbers, which shall be determined by adding to the whole Number of free Persons, including those bound to Service for a Term of Years, and excluding Indians not taxed, three fifths of all other Persons. The actual Enumeration shall be made within three Years after the first Meeting of the Congress of the United States, and within every subsequent Term of ten Years, in such Manner as they shall by Law direct. The Number of Representatives shall not exceed one for every thirty Thousand, but each State shall have at Least one Representative; and until such enumeration shall be made, the State of New Hampshire shall be entitled to chuse three, Massachusetts eight, Rhode-Island and Providence Plantations one, Connecticut five, New-York six, New Jersey four, Pennsylvania eight, Delaware one, Maryland six, Virginia ten, North Carolina five, South Carolina five, and Georgia three.

When vacancies happen in the Representation from any State, the Executive Authority thereof shall issue Writs of Election to fill such Vacancies.

The House of Representatives shall chuse their Speaker and other Officers; and shall have the sole Power of Impeachment.

Section. 3. The Senate of the United States shall be composed of two Senators from each State, chosen by the Legislature thereof, for six Years; and each Senator shall have one Vote.

Immediately after they shall be assembled in Consequence of the first Election, they shall be divided as equally as may be into three Classes. The Seats of the Senators of the first Class shall be vacated at the Expiration of the second Year, of the second Class at the Expiration of the fourth Year, and of the third Class at the Expiration of the sixth Year, so that one third may be chosen every second Year; and if Vacancies happen by Resignation, or otherwise, during the Recess of the Legislature of any State, the Executive thereof may make temporary Appointments until the next Meeting of the Legislature, which shall then fill such Vacancies.

No Person shall be a Senator who shall not have attained to the Age of thirty Years, and been nine Years a Citizen of the United States, and who shall not, when elected, be an Inhabitant of that State for which he shall be chosen.

The Vice President of the United States shall be President of the Senate, but shall have no Vote, unless they be equally divided.

The Senate shall chuse their other Officers, and also a President pro tempore, in the Absence of the Vice President, or when he shall exercise the Office of President of the United States.

The Senate shall have the sole Power to try all Impeachments. When sitting for that Purpose, they shall be on Oath or Affirmation. When the President of the United States is tried, the Chief Justice shall preside: And no Person shall be convicted without the Concurrence of two thirds of the Members present.

Judgment in Cases of Impeachment shall not extend further than to removal from Office, and disqualification to hold and enjoy any Office of honor, Trust or Profit under the United States: but the Party convicted shall nevertheless be liable and subject to Indictment, Trial, Judgment and Punishment, according to Law.

Section. 4. The Times, Places and Manner of holding Elections for Senators and Representatives, shall be prescribed in each State by the Legislature thereof; but the Congress may at any time by Law make or alter such Regulations, except as to the Places of chusing Senators.

The Congress shall assemble at least once in every Year, and such Meeting shall be on the first Monday in December [Modified by Amendment XX], unless they shall by Law appoint a different Day.

Section. 5. Each House shall be the Judge of the Elections, Returns and Qualifications of its own Members, and a Majority of each shall constitute a Quorum to do Business; but a smaller Number may adjourn from day to day, and may be authorized to compel the Attendance of absent Members, in such Manner, and under such Penalties as each House may provide.

Each House may determine the Rules of its Proceedings, punish its Members for disorderly Behaviour, and, with the Concurrence of two thirds, expel a Member.

Each House shall keep a Journal of its Proceedings, and from time to time publish the same, excepting such Parts as may in their Judgment require Secrecy; and the Yeas and Nays of the Members of either House on any question shall, at the Desire of one fifth of those Present, be entered on the Journal.

Neither House, during the Session of Congress, shall, without the Consent of the other, adjourn for more than three days, nor to any other Place than that in which the two Houses shall be sitting.

Section. 6. The Senators and Representatives shall receive a Compensation for their Services, to be ascertained by Law, and paid out of the Treasury of the United States. They shall in all Cases, except Treason, Felony and Breach of the Peace, be privileged from Arrest during their Attendance at the Session of their respective Houses, and in going to and returning from the same; and for any Speech or Debate in either House, they shall not be questioned in any other Place.

No Senator or Representative shall, during the Time for which he was elected, be appointed to any civil Office under the Authority of the United States, which shall have been created, or the Emoluments whereof shall have been encreased during such time; and no Person holding any Office under the United States, shall be a Member of either House during his Continuance in Office.

Section. 7. All Bills for raising Revenue shall originate in the House of Representatives; but the Senate may propose or concur with Amendments as on other Bills.

Every Bill which shall have passed the House of Representatives and the Senate, shall, before it become a Law, be presented to the President of the United States: If he approve he shall sign it, but if not he shall return it, with his Objections to that House in which it shall have originated, who shall enter the Objections at large on their Journal, and proceed to reconsider it. If after such Reconsideration two thirds of that House shall agree to pass the Bill, it shall be sent, together with the Objections, to the other House, by which it shall likewise be reconsidered, and if approved by two thirds of that House, it shall become a Law. But in all such Cases the Votes of both Houses shall be determined by yeas and Nays, and the Names of the Persons voting for and against the Bill shall be entered on the Journal of each House respectively. If any Bill shall not be returned by the President within ten Days (Sundays excepted) after it shall have been presented to him, the Same shall be a Law, in like Manner as if he had signed it, unless the Congress by their Adjournment prevent its Return, in which Case it shall not be a Law.

Every Order, Resolution, or Vote to which the Concurrence of the Senate and House of Representatives may be necessary (except on a question of Adjournment) shall be presented to the President of the United States; and before the Same shall take Effect, shall be approved by him, or being disapproved by him, shall be repassed by two thirds of the Senate and House of Representatives, according to the Rules and Limitations prescribed in the Case of a Bill.

Section. 8. The Congress shall have Power To lay and collect Taxes, Duties, Imposts and Excises, to pay the Debts and provide for the common Defense and general Welfare of the United States; but all Duties, Imposts and Excises shall be uniform throughout the United States;

To borrow Money on the credit of the United States;

To regulate Commerce with foreign Nations, and among the several States, and with the Indian Tribes;

To establish an uniform Rule of Naturalization, and uniform Laws on the subject of Bankruptcies throughout the United States;

<u>To coin Money, regulate the Value thereof, and of foreign Coin, and fix the Standard of Weights and Measures;</u>

<u>To provide for the Punishment of counterfeiting the Securities and current Coin of the United States;</u>

To establish Post Offices and post Roads;

To promote the Progress of Science and useful Arts, by securing for limited Times to Authors and Inventors the exclusive Right to their respective Writings and Discoveries;

To constitute Tribunals inferior to the supreme Court;

To define and punish Piracies and Felonies committed on the high Seas, and Offences against the Law of Nations;

To declare War, grant Letters of Marque and Reprisal, and make Rules concerning Captures on Land and Water;

To raise and support Armies, but no Appropriation of Money to that Use shall be for a longer Term than two Years;

To provide and maintain a Navy;

To make Rules for the Government and Regulation of the land and naval Forces;

To provide for calling forth the Militia to execute the Laws of the Union, suppress Insurrections and repel Invasions;

To provide for organizing, arming, and disciplining, the Militia, and for governing such Part of them as may be employed in the Service of the United States, reserving to the States respectively, the Appointment of the Officers, and the Authority of training the Militia according to the discipline prescribed by Congress;

To exercise exclusive Legislation in all Cases whatsoever, over such District (not exceeding ten Miles square) as may, by Cession of particular States, and the Acceptance of Congress, become the Seat of the Government of the United States, and to exercise like Authority over all Places purchased by the Consent of the Legislature of the State in which the Same shall be, for the Erection of Forts, Magazines, Arsenals, dock-Yards, and other needful Buildings; --And

To make all Laws which shall be necessary and proper for carrying into Execution the foregoing Powers, and all other Powers vested by this Constitution in the Government of the United States, or in any Department or Officer thereof.

Section. 9. The Migration or Importation of such Persons as any of the States now existing shall think proper to admit, shall not be prohibited by the Congress prior to the Year one thousand eight hundred and eight, but a Tax or duty may be imposed on such Importation, not exceeding ten dollars for each Person.

The Privilege of the Writ of Habeas Corpus shall not be suspended, unless when in Cases of Rebellion or Invasion the public Safety may require it.

No Bill of Attainder or ex post facto Law shall be passed.

No Capitation, or other direct, Tax shall be laid, unless in Proportion to the Census or Enumeration herein before directed to be taken.

No Tax or Duty shall be laid on Articles exported from any State.

No Preference shall be given by any Regulation of Commerce or Revenue to the Ports of one State over those of another; nor shall Vessels bound to, or from, one State, be obliged to enter, clear, or pay Duties in another.

No Money shall be drawn from the Treasury, but in Consequence of Appropriations made by Law; and a regular Statement and Account of the Receipts and Expenditures of all public Money shall be published from time to time.

No Title of Nobility shall be granted by the United States: And no Person holding any Office of Profit or Trust under them, shall, without the Consent of the Congress, accept of any present, Emolument, Office, or Title, of any kind whatever, from any King, Prince, or foreign State.

Section. 10. <u>No State shall</u> enter into any Treaty, Alliance, or Confederation; grant Letters of Marque and Reprisal; <u>coin Money; emit Bills of Credit; make any Thing but gold and silver Coin a Tender in Payment of Debts</u>; pass any Bill of Attainder, ex post facto Law, or Law impairing the Obligation of Contracts, or grant any Title of Nobility.

No State shall, without the Consent of the Congress, lay any Imposts or Duties on Imports or Exports, except what may be absolutely necessary for executing it's inspection Laws; and the net Produce of all Duties and Imposts, laid by any State on Imports or Exports, shall be for the Use of the Treasury of the United States;

and all such Laws shall be subject to the Revision and Controul of the Congress.

No State shall, without the Consent of Congress, lay any Duty of Tonnage, keep Troops, or Ships of War in time of Peace, enter into any Agreement or Compact with another State, or with a foreign Power, or engage in War, unless actually invaded, or in such imminent Danger as will not admit of delay.

Article. II.

Section. 1. The executive Power shall be vested in a President of the United States of America. He shall hold his Office during the Term of four Years, and, together with the Vice President, chosen for the same Term, be elected, as follows:

Each State shall appoint, in such Manner as the Legislature thereof may direct, a Number of Electors, equal to the whole Number of Senators and Representatives to which the State may be entitled in the Congress: but no Senator or Representative, or Person holding an Office of Trust or Profit under the United States, shall be appointed an Elector.

The Electors shall meet in their respective States, and vote by Ballot for two Persons, of whom one at least shall not be an Inhabitant of the same State with themselves. And they shall make a List of all the Persons voted for, and of the Number of Votes for each; which List they shall sign and certify, and transmit sealed to the Seat of the Government of the United States, directed to the President of the Senate. The President of the Senate shall, in the Presence of the Senate and House of Representatives, open all the Certificates, and the Votes shall then be counted. The Person having the greatest Number of Votes shall be the President, if such Number be a Majority of the whole Number of Electors appointed; and if there be more than one who have such Majority, and have an equal Number of Votes, then the House of Representatives shall immediately chuse by Ballot one of them for President; and if no Person have a Majority, then from the five highest on the List the said House shall in like Manner chuse the President. But in chusing the President, the Votes shall be taken by States, the Representation from each State having one Vote; a quorum for this Purpose shall consist of a Member or Members from two thirds of the States, and a Majority of all the States shall be necessary to a Choice. In every Case, after the Choice of the President, the Person having the greatest Number of Votes of the Electors shall be the Vice President. But if there should remain two or more who have equal Votes, the Senate shall chuse from them by Ballot the Vice President.

The Congress may determine the Time of chusing the Electors, and the Day on which they shall give their Votes; which Day shall be the same throughout the United States.

No Person except a natural born Citizen, or a Citizen of the United States, at the time of the Adoption of this Constitution, shall be eligible to the Office of President; neither shall any Person be eligible to that Office who shall not have attained to the Age of thirty five Years, and been fourteen Years a Resident within the United States.

In Case of the Removal of the President from Office, or of his Death, Resignation, or Inability to discharge the Powers and Duties of the said Office, the Same shall devolve on the Vice President, and the Congress may by Law provide for the Case of Removal, Death, Resignation or Inability, both of the President and Vice President, declaring what Officer shall then act as President, and such Officer shall act accordingly, until the Disability be removed, or a President shall be elected.

The President shall, at stated Times, receive for his Services, a Compensation, which shall neither be increased nor diminished during the Period for which he shall have been elected, and he shall not receive within that Period any other Emolument from the United States, or any of them.

Before he enter on the Execution of his Office, he shall take the following Oath or Affirmation: --"I do solemnly swear (or affirm) that I will faithfully execute the Office of President of the United States, and will to the best of my Ability, preserve, protect and defend the Constitution of the United States."

Section. 2. The President shall be Commander in Chief of the Army and Navy of the United States, and of the Militia of the several States, when called into the actual Service of the United States; he may require the Opinion, in writing, of the principal Officer in each of the executive Departments, upon any Subject relating to the Duties of their respective Offices, and he shall have Power to grant Reprieves and Pardons for Offences against the United States, except in Cases of Impeachment.

He shall have Power, by and with the Advice and Consent of the Senate, to make Treaties, provided two thirds of the Senators present concur; and he shall nominate, and by and with the Advice and Consent of the Senate, shall appoint Ambassadors, other public Ministers and Consuls, Judges of the supreme Court, and all other Officers of the United States, whose Appointments are not herein otherwise provided for, and which shall be established by Law: but the Congress may by Law vest the Appointment of such inferior Officers, as they think proper, in the President alone, in the Courts of Law, or in the Heads of Departments.

The President shall have Power to fill up all Vacancies that may happen during the Recess of the Senate, by granting Commissions which shall expire at the End of their next Session.

Section. 3. He shall from time to time give to the Congress Information of the State of the Union, and recommend to their Consideration such Measures as he shall judge necessary and expedient; he may, on extraordinary Occasions, convene both Houses, or either of them, and in Case of Disagreement between them, with Respect to the Time of Adjournment, he may adjourn them to such Time as he shall think proper; he shall receive Ambassadors and other public Ministers; he shall take Care that the Laws be faithfully executed, and shall Commission all the Officers of the United States.

Section. 4. The President, Vice President and all civil Officers of the United States, shall be removed from Office on Impeachment for, and Conviction of, Treason, Bribery, or other high Crimes and Misdemeanors.

Article. III.

Section. 1. The judicial Power of the United States shall be vested in one supreme Court, and in such inferior Courts as the Congress may from time to time ordain and establish. The Judges, both of the supreme and inferior Courts, shall hold their Offices during good Behaviour, and shall, at stated Times, receive for their Services a Compensation, which shall not be diminished during their Continuance in Office.

Section. 2. The judicial Power shall extend to all Cases, in Law and Equity, arising under this Constitution, the Laws of the United States, and Treaties made, or which shall be made, under their Authority; --to all Cases affecting Ambassadors, other public Ministers and Consuls; --to all Cases of admiralty and maritime Jurisdiction; --to Controversies to which the United States shall be a Party; --to Controversies between two or more States; --between a State and Citizens of another State; --between Citizens of different States; --between Citizens of the same State claiming Lands under Grants of different States, and between a State, or the Citizens thereof, and foreign States, Citizens or Subjects.

In all Cases affecting Ambassadors, other public Ministers and Consuls, and those in which a State shall be Party, the supreme Court shall have original Jurisdiction. In all the other Cases before mentioned, the supreme Court shall have appellate Jurisdiction, both as to Law and Fact, with such Exceptions, and under such Regulations as the Congress shall make.

The Trial of all Crimes, except in Cases of Impeachment, shall be by Jury; and such Trial shall be held in the State where the said Crimes shall have been committed; but when not committed within any State, the Trial shall be at such Place or Places as the Congress may by Law have directed.

Section. 3. Treason against the United States shall consist only in levying War against them, or in adhering to their Enemies, giving them Aid and Comfort. No Person shall be convicted of Treason unless on the Testimony of two Witnesses to the same overt Act, or on Confession in open Court.

The Congress shall have Power to declare the Punishment of Treason, but no Attainder of Treason shall work Corruption of Blood, or Forfeiture except during the Life of the Person attainted.

Article. IV.

Section. 1. Full Faith and Credit shall be given in each State to the public Acts, Records, and judicial Proceedings of every other State. And the Congress may by general Laws prescribe the Manner in which such Acts, Records and Proceedings shall be proved, and the Effect thereof.

Section. 2. The Citizens of each State shall be entitled to all Privileges and Immunities of Citizens in the several States.

A Person charged in any State with Treason, Felony, or other Crime, who shall flee from Justice, and be found in another State, shall on Demand of the executive Authority of the State from which he fled, be delivered up, to be removed to the State having Jurisdiction of the Crime.

No Person held to Service or Labour in one State, under the Laws thereof, escaping into another, shall, in Consequence of any Law or Regulation therein, be discharged from such Service or Labour, but shall be delivered up on Claim of the Party to whom such Service or Labour may be due.

Section. 3. New States may be admitted by the Congress into this Union; but no new State shall be formed or erected within the Jurisdiction of any other State; nor any State be formed by the Junction of two or more States, or Parts of States, without the Consent of the Legislatures of the States concerned as well as of the Congress.

The Congress shall have Power to dispose of and make all needful Rules and Regulations respecting the Territory or other Property belonging to the United States; and nothing in this Constitution shall be so construed as to Prejudice any Claims of the United States, or of any particular State.

Section. 4. The United States shall guarantee to every State in this Union a Republican Form of Government, and shall protect each of them against Invasion; and on Application of the Legislature, or of the Executive (when the Legislature cannot be convened), against domestic Violence.

Article. V.

The Congress, whenever two thirds of both Houses shall deem it necessary, shall propose Amendments to this Constitution, or, on the Application of the Legislatures of two thirds of the several States, shall call a Convention for proposing Amendments, which, in either Case, shall be valid to all Intents and Purposes, as Part of this Constitution, when ratified by the Legislatures of three fourths of the several States, or by Conventions in three fourths thereof, as the one or the other Mode of Ratification may be proposed by the Congress; Provided that no Amendment which may be made prior to the Year One thousand eight hundred and eight shall in any Manner affect the first and fourth Clauses in the Ninth Section of the first Article; and that no State, without its Consent, shall be deprived of its equal Suffrage in the Senate.

Article. VI.

All Debts contracted and Engagements entered into, before the Adoption of this Constitution, shall be as valid against the United States under this Constitution, as under the Confederation.

This Constitution, and the Laws of the United States which shall be made in Pursuance thereof; and all Treaties made, or which shall be made, under the Authority of the United States, shall be the supreme Law of the Land; and the Judges in every State shall be bound thereby, any Thing in the Constitution or Laws of any State to the Contrary notwithstanding.

The Senators and Representatives before mentioned, and the Members of the several State Legislatures, and all executive and judicial Officers, both of the United States and of the several States, shall be bound by Oath or Affirmation, to support this Constitution; but no religious Test shall ever be required as a Qualification to any Office or public Trust under the United States.

Article. VII.

The Ratification of the Conventions of nine States, shall be sufficient for the Establishment of this Constitution between the States so ratifying the Same.

The Word, "the," being interlined between the seventh and eighth Lines of the first Page, The Word "Thirty" being partly written on an Erazure in the fifteenth Line of the first Page, The Words "is tried" being interlined between the thirty second and thirty third Lines of the first Page and the Word "the" being interlined between the forty third and forty fourth Line of the second Page.

Attest William Jackson
Secretary

Done in Convention by the Unanimous Consent of the States present the Seventeenth Day of September in the Year of our Lord one thousand seven hundred and Eighty seven and of the Independence of the United States of America the Twelfth In witness whereof We have hereunto subscribed our Names,

Go. Washington--Presidt.
and deputy from Virginia

New Hampshire {
  John Langdon
  Nicholas Gilman

Massachusetts {
  Nathaniel Gorham
  Rufus King

Connecticut {
  Wm. Saml. Johnson
  Roger Sherman

New York . . . .
  Alexander Hamilton

New Jersey {
  Wil: Livingston
  David Brearley.
  Wm. Paterson.
  Jona: Dayton

Pennsylvania {
  B Franklin
  Thomas Mifflin
  Robt Morris
  Geo. Clymer
  Thos. Fitz Simons
  Jared Ingersoll
  James Wilson
  Gouv Morris

Delaware {
  Geo: Read
  Gunning Bedford jun
  John Dickinson
  Richard Bassett
  Jaco: Broom

Maryland {
  James Mchenry
  Dan of St Thos. Jenifer
  Danl Carroll

Virginia {
  John Blair
  James Madison

North Carolina {
  Wm. Blount
  Richd. Dobbs Spaight
  Hu Williamson
  J. Rutledge

South Carolina {
  Charles Cotesworth Pinckney
  Charles Pinckney
  Pierce Butler

Georgia {
  William Few
  Abr Baldwin

In Convention Monday, September 17th, 1787

Present The States of
New Hampshire, Massachusetts, Connecticut, Mr. Hamilton from
New York, New Jersey, Pennsylvania, Delaware, Maryland, Virginia,
North Carolina, South Carolina and Georgia.

Resolved,

That the preceeding Constitution be laid before the United States in
Congress assembled, and that it is the Opinion of this Convention,
that it should afterwards be submitted to a Convention of Delegates,
chosen in each State by the People thereof, under the
Recommendation of its Legislature, for their Assent and
Ratification; and that each Convention assenting to,and ratifying the
Same, should give Notice thereof to the United States in Congress
assembled. Resolved, That it is the Opinion of this Convention, that
as soon as the Conventions of nine States shall have ratified this
Constitution, the United States in Congress assembled should fix a
Day on which Electors should be appointed by the States which
have ratified the same, and a Day on which the Electors should
assemble to vote for the President, and the Time and Place for com-
mencing Proceedings under this Constitution. That after such
Publication the Electors should be appointed, and the Senators and
Representatives elected: That the Electors should meet on the Day
fixed for the Election of the President, and should transmit their
Votes certified, signed, sealed and directed, as the Constitution
requires, to the Secretary of the United States in Congress assembled, that the Senators and Representatives should convene at the
Time and Place assigned; that the Senators should appoint a
President of the Senate, for the sole purpose of receiving, opening
and counting the Votes for President; and, that after he shall be
chosen, the Congress, together with the President, should, without
Delay, proceed to execute this Constitution.

By the Unanimous Order of the Convention

Go. Washington--Presidt.
W. Jackson Secretary.

The Bill of Rights

The conventions of a number of the States having at the time of their adopting the Constitution, expressed a desire, in order to prevent misconstruction or abuse of its powers, that further declaratory and restrictive clauses should be added.

Article the first [Not Ratified]
After the first enumeration required by the first article of the Constitution, there shall be one Representative for every thirty thousand, until the number shall amount to one hundred, after which the proportion shall be so regulated by Congress, that there shall be not less than one hundred Representatives, nor less than one Representative for every forty thousand persons, until the number of Representatives shall amount to two hundred; after which the proportion shall be so regulated by Congress, that there shall not be less than two hundred Representatives, nor more than one Representative for every fifty thousand persons.

Article the second [Amendment XXVII - Ratified 1992]
No law, varying the compensation for the services of the Senators and Representatives, shall take effect, until an election of Representatives shall have intervened.

Article the third [Amendment I]

Congress shall make no law respecting an establishment of religion, or prohibiting the free exercise thereof; or abridging the freedom of speech, or of the press; or the right of the people peaceably to assemble, and to petition the Government for a redress of grievances.

Article the fourth [Amendment II]

A well regulated Militia, being necessary to the security of a free State, the right of the people to keep and bear Arms, shall not be infringed.

Article the fifth [Amendment III]

No Soldier shall, in time of peace be quartered in any house, without the consent of the Owner, nor in time of war, but in a manner to be prescribed by law.

Article the sixth [Amendment IV]

The right of the people to be secure in their persons, houses, papers, and effects, against unreasonable searches and seizures, shall not be violated, and no Warrants shall issue, but upon probable cause, supported by Oath or affirmation, and particularly describing the place to be searched, and the persons or things to be seized.

Article the seventh [Amendment V]

No person shall be held to answer for a capital, or otherwise infamous crime, unless on a presentment or indictment of a Grand Jury, except in cases arising in the land or naval forces, or in the Militia, when in actual service in time of War or public danger; nor shall any person be subject for the same offence to be twice put in jeopardy of life or limb; nor shall be compelled in any criminal case to be a witness against himself, nor be deprived of life, liberty, or property, without due process of law; nor shall private property be taken for public use, without just compensation.

Article the eighth [Amendment VI]

In all criminal prosecutions, the accused shall enjoy the right to a speedy and public trial, by an impartial jury of the State and district wherein the crime shall have been committed, which district shall have been previously ascertained by law, and to be informed of the nature and cause of the accusation; to be confronted with the witnesses against him; to have compulsory process for obtaining witnesses in his favor, and to have the Assistance of Counsel for his defense.

Article the ninth [Amendment VII]
In Suits at common law, where the value in controversy shall exceed twenty dollars, the right of trial by jury shall be preserved, and no fact tried by a jury, shall be otherwise re-examined in any Court of the United States, than according to the rules of the common law.

Article the tenth [Amendment VIII]
Excessive bail shall not be required, nor excessive fines imposed, nor cruel and unusual punishments inflicted.

Article the eleventh [Amendment IX]
The enumeration in the Constitution, of certain rights, shall not be construed to deny or disparage others retained by the people.

Article the twelfth [Amendment X]
The powers not delegated to the United States by the Constitution, nor prohibited by it to the States, are reserved to the States respectively, or to the people.

Additional Amendments to the Constitution

ARTICLES in addition to, and Amendment of, the Constitution of the United States of America, proposed by Congress, and ratified by the Legislatures of the several States, pursuant to the fifth Article of the original Constitution

[Article. XI.]
[Proposed 1794; Ratified 1798]

The Judicial power of the United States shall not be construed to extend to any suit in law or equity, commenced or prosecuted against one of the United States by Citizens of another State, or by Citizens or Subjects of any Foreign State.

[Article. XII.]
[Proposed 1803; Ratified 1804]

The Electors shall meet in their respective states, and vote by ballot for President and Vice-President, one of whom, at least, shall not be an inhabitant of the same state with themselves; they shall name in their ballots the person voted for as President, and in distinct ballots the person voted for as Vice-President, and they shall make distinct lists of all persons voted for as President, and of all persons voted for as Vice-President, and of the number of votes for each, which lists they shall sign and certify, and transmit sealed to the seat of the government of the United States, directed to the President of the Senate;-- The President of the Senate shall, in the presence of the Senate and House of Representatives, open all the certificates and the votes shall then be counted;--The person having the greatest number of votes for President, shall be the President, if such number be a majority of the whole number of Electors appointed; and if no person have such majority, then from the persons having the highest numbers not exceeding three on the list of those voted for as President, the House of Representatives shall choose immediately, by ballot, the President. But in choosing the President, the votes shall be taken by states, the representation from each state having one vote; a quorum for this purpose shall consist of a member or members from two-thirds of the states, and a majority of all the states shall be necessary to a choice. And if the House of Representatives shall not choose a President whenever the right of choice shall devolve upon them, before the fourth day of March next following, then the Vice-President shall act as President, as in the case of the death or other constitutional disability of the President.-- The person having the greatest number of votes as Vice-President, shall be the Vice-President, if such number be a majority of the whole number of Electors appointed, and if no person have a majority, then from the two highest numbers on the list, the Senate shall choose the Vice-President; a quorum for the purpose shall consist of two-thirds of the whole number of Senators, and a majority of the whole number shall be necessary to a choice. But no person constitutionally ineligible to the office of President shall be eligible to that of Vice-President of the United States.

[Contested Article.]
[Proposed 1810; Probably Ratified 1819]

If any Citizen of the United States shall accept, claim, receive or retain any Title of Nobility or Honour, or shall, without the Consent of Congress, accept and retain any present, Pension, Office or Emolument of any kind whatever, from any Emperor, King, Prince or foreign Power, such Person shall cease to be a Citizen of the United States, and shall be incapable of holding any Office of Trust or Profit under them, or either of them.

[Unratified Article.]
[Proposed 1861; Signed by President Lincoln; Unratified]

Article Thirteen.

No amendment shall be made to the Constitution which will authorize or give to Congress the power to abolish or interfere, within any State, with the domestic institutions thereof, including that of persons held to labor or service by the laws of said State.

Article. XIII.
[Proposed 1865; Ratified 1865]

Section. 1. Neither slavery nor involuntary servitude, except as a punishment for crime whereof the party shall have been duly convicted, shall exist within the United States, or any place subject to their jurisdiction.

Section. 2. Congress shall have power to enforce this article by appropriate legislation.

Article. XIV.
[Proposed 1866; Ratified Under Duress 1868]

Section. 1. All persons born or naturalized in the United States, and subject to the jurisdiction thereof, are citizens of the United States and of the State wherein they reside. No State shall make or enforce any law which shall abridge the privileges or immunities of citizens of the United States; nor shall any State deprive any person of life, liberty, or property, without due process of law; nor deny to any person within its jurisdiction the equal protection of the laws.

Section. 2. Representatives shall be apportioned among the several States according to their respective numbers, counting the whole number of persons in each State, excluding Indians not taxed. But when the right to vote at any election for the choice of electors for President and Vice President of the United States, Representatives in Congress, the Executive and Judicial officers of a State, or the members of the Legislature thereof, is denied to any of the male inhabitants of such State, being twenty-one years of age, and citizens of the United States, or in any way abridged, except for participation in rebellion, or other crime, the basis of representation therein shall be reduced in the proportion which the number of such male citizens shall bear to the whole number of male citizens twenty-one years of age in such State.

Section. 3. No person shall be a Senator or Representative in Congress, or elector of President and Vice President, or hold any office, civil or military, under the United States, or under any State, who, having previously taken an oath, as a member of Congress, or as an officer of the United States, or as a member of any State legislature, or as an executive or judicial officer of any State, to support the Constitution of the United States, shall have engaged in insurrection or rebellion against the same, or given aid or comfort to the enemies thereof. But Congress may by a vote of two-thirds of each

House, remove such disability.

Section. 4. The validity of the public debt of the United States, authorized by law, including debts incurred for payment of pensions and bounties for services in suppressing insurrection or rebellion, shall not be questioned. But neither the United States nor any State shall assume or pay any debt or obligation incurred in aid of insurrection or rebellion against the United States, or any claim for the loss or emancipation of any slave; but all such debts, obligations and claims shall be held illegal and void.

Section. 5. The Congress shall have power to enforce, by appropriate legislation, the provisions of this article.

Article. XV.
[Proposed 1869; Ratified 1870]

Section. 1. The right of citizens of the United States to vote shall not be denied or abridged by the United States or by any State on account of race, color, or previous condition of servitude.

Section. 2. The Congress shall have power to enforce this article by appropriate legislation.

Article. XVI.
[Proposed 1909; Questionably Ratified 1913]

The Congress shall have power to lay and collect taxes on incomes, from whatever source derived, without apportionment among the several States, and without regard to any census or enumeration.

[Article. XVII.]
[Proposed 1912; Ratified 1913; Possibly Unconstitutional
(See Article V, Clause 3 of the Constitution)]

The Senate of the United States shall be composed of two Senators from each State, elected by the people thereof, for six years; and each Senator shall have one vote. The electors in each State shall have the qualifications requisite for electors of the most numerous branch of the State legislatures.

When vacancies happen in the representation of any State in the Senate, the executive authority of such State shall issue writs of election to fill such vacancies: Provided, That the legislature of any State may empower the executive thereof to make temporary appointments until the people fill the vacancies by election as the legislature may direct.

This amendment shall not be so construed as to affect the election or term of any Senator chosen before it becomes valid as part of the Constitution.

Article. [XVIII.]
[Proposed 1917; Ratified 1919; Repealed 1933
(See Amendment XXI, Section 1A)]

Section. 1. After one year from the ratification of this article the manufacture, sale, or transportation of intoxicating liquors within, the importation thereof into, or the exportation thereof from the United States and all territory subject to the jurisdiction thereof for beverage purposes is hereby prohibited.

Section. 2. The Congress and the several States shall have concurrent power to enforce this article by appropriate legislation.

Section. 3. This article shall be inoperative unless it shall have been ratified as an amendment to the Constitution by the legislatures of the several States, as provided in the Constitution, within seven years from the date of the submission hereof to the States by the Congress.

Article. [XIX.]
[Proposed 1919; Ratified 1920]

The right of citizens of the United States to vote shall not be denied or abridged by the United States or by any State on account of sex. Congress shall have power to enforce this article by appropriate legislation.

[Unratified Article.]
[Proposed 1926; Unratified]

Section. 1. The Congress shall have power to limit, regulate, and prohibit the labor of persons under eighteen years of age.

Section. 2. The power of the several States is unimpaired by this article except that the operation of State laws shall be suspended to the extent necessary to give effect to legislation enacted by the Congress.

Article. [XX.]
[Proposed 1932; Ratified 1933]

Section. 1. The terms of the President and Vice President shall end at noon on the 20th day of January, and the terms of Senators and Representatives at noon on the 3d day of January, of the years in which such terms would have ended if this article had not been ratified; and the terms of their successors shall then begin.

Section. 2. The Congress shall assemble at least once in every year, and such meeting shall begin at noon on the 3d day of January, unless they shall by law appoint a different day.

Section. 3. If, at the time fixed for the beginning of the term of the President, the President elect shall have died, the Vice President elect shall become President. If a President shall not have been chosen before the time fixed for the beginning of his term, or if the President elect shall have failed to qualify, then the Vice President elect shall act as President until a President shall have qualified; and the Congress may by law provide for the case wherein neither a President elect nor a Vice President elect shall have qualified, declaring who shall then act as President, or the manner in which one who is to act shall be selected, and such person shall act accordingly until a President or Vice President shall have qualified.

Section. 4. The Congress may by law provide for the case of the death of any of the persons from whom the House of Representatives may choose a President whenever the right of choice shall have devolved upon them, and for the case of the death of any of the persons from whom the Senate may choose a Vice President whenever the right of choice shall have devolved upon them.

Section. 5. Sections 1 and 2 shall take effect on the 15th day of October following the ratification of this article.

Section. 6. This article shall be inoperative unless it shall have been ratified as an amendment to the Constitution by the legislatures of three-fourths of the several States within seven years from the date of its submission.

Article. [XXI.]
[Proposed 1933; Ratified 1933]

Section. 1. The eighteenth article of amendment to the Constitution of the United States is hereby repealed.

Section. 2. The transportation or importation into any State, Territory, or possession of the United States for delivery or use therein of intoxicating liquors, in violation of the laws thereof, is hereby prohibited.

Section. 3. This article shall be inoperative unless it shall have been ratified as an amendment to the Constitution by conventions in the several States, as provided in the Constitution, within seven years from the date of the submission hereof to the States by the Congress.

Article. [XXII.]
[Proposed 1947; Ratified 1951]

Section. 1. No person shall be elected to the office of the President more than twice, and no person who has held the office of President, or acted as President, for more than two years of a term to which some other person was elected President shall be elected to the office of the President more than once. But this Article shall not apply to any person holding the office of President when this Article was proposed by the Congress, and shall not prevent any person who may be holding the office of President, or acting as

President, during the term within which this Article becomes operative from holding the office of President or acting as President during the remainder of such term.

Section. 2. This article shall be inoperative unless it shall have been ratified as an amendment to the Constitution by the legislatures of three-fourths of the several States within seven years from the date of its submission to the States by the Congress.

Article. [XXIII.]
[Proposed 1960; Ratified 1961]

Section. 1. The District constituting the seat of Government of the United States shall appoint in such manner as the Congress may direct:

A number of electors of President and Vice President equal to the whole number of Senators and Representatives in Congress to which the District would be entitled if it were a State, but in no event more than the least populous State; they shall be in addition to those appointed by the States, but they shall be considered, for the purposes of the election of President and Vice President, to be electors appointed by a State; and they shall meet in the District and perform such duties as provided by the twelfth article of amendment.

Section. 2. The Congress shall have power to enforce this article by appropriate legislation.

Article. [XXIV.]
[Proposed 1962; Ratified 1964]

Section. 1. The right of citizens of the United States to vote in any primary or other election for President or Vice President, for electors for President or Vice President, or for Senator or Representative in Congress, shall not be denied or abridged by the United States or any State by reason of failure to pay any poll tax or other tax.

Section. 2. The Congress shall have power to enforce this article by appropriate legislation.

Article. [XXV.]
[Proposed 1965; Ratified 1967]

Section. 1. In case of the removal of the President from office or of his death or resignation, the Vice President shall become President.

Section. 2. Whenever there is a vacancy in the office of the Vice President, the President shall nominate a Vice President who shall take office upon confirmation by a majority vote of both Houses of Congress.

Section. 3. Whenever the President transmits to the President pro tempore of the Senate and the Speaker of the House of Representatives his written declaration that he is unable to discharge the powers and duties of his office, and until he transmits to them a written declaration to the contrary, such powers and duties shall be discharged by the Vice President as Acting President.

Section. 4. Whenever the Vice President and a majority of either the principal officers of the executive departments or of such other body as Congress may by law provide, transmit to the President pro tempore of the Senate and the Speaker of the House of Representatives their written declaration that the President is unable to discharge the powers and duties of his office, the Vice President shall immediately assume the powers and duties of the office as Acting President.

Thereafter, when the President transmits to the President pro tempore of the Senate and the Speaker of the House of Representatives his written declaration that no inability exists, he shall resume the powers and duties of his office unless the Vice President and a majority of either the principal officers of the executive department or of such other body as Congress may by law provide, transmit within four days to the President pro tempore of the Senate and the Speaker of the House of Representatives their written declaration that the President is unable to discharge the powers and duties of his office. Thereupon Congress shall decide the issue, assembling within forty-eight hours for that purpose if not in session. If the Congress, within twenty-one days after receipt of the latter written declaration, or, if Congress is not in session, within twenty-one days after Congress is required to assemble, determines by two-thirds vote of both Houses that the President is unable to discharge the powers and duties of his office, the Vice

President shall continue to discharge the same as Acting President; otherwise, the President shall resume the powers and duties of his office.

Article. [XXVI.]
[Proposed 1971; Ratified 1971]

Section. 1. The right of citizens of the United States, who are eighteen years of age or older, to vote shall not be denied or abridged by the United States or by any State on account of age.

Section. 2. The Congress shall have power to enforce this article by appropriate legislation.

[Inoperative Article.]
Proposed 1978; Expired Unratified 1985]

Section. 1. For purposes of representation in the Congress, election of the President and Vice President, and article V of this Constitution, the District constituting the seat of government of the United States shall be treated as though it were a State.

Section. 2. The exercise of the rights and powers conferred under this article shall be by the people of the District constituting the seat of government, and as shall be provided by the Congress.

Section. 3. The twenty-third article of amendment to the Constitution of the United States is hereby repealed.

Section. 4. This article shall be inoperative, unless it shall have been ratified as an amendment to the Constitution by the legislatures of three-fourths of the several States within seven years from the date of its submission.

Article. [XXVII.]
[Proposed 1789; Ratified 1992;
Second of twelve Articles comprising the Bill of Rights]

No law, varying the compensation for the services of the Senators and Representatives, shall take effect, until an election of Representatives shall have intervened.

SOURCE: www..constitution.org. Recommended reading: *Citizen's Rule Book*, ed. Webster Adams, Whitten Printers, 1001 S. 5th St, Phoenix, AZ.

III. Bibliography

I. History of U.S. Rare Coins

Breen, Walter. *Walter Breen's Complete Encyclopedia of U.S. and Colonial Coins,* FCI Press, Doubleday, NY, NY, 1988

Miller, Wayne. *The Morgan and Peace Dollar Textbook*, Adam Smith Publishing, Metairie, LA, 1983

Swiatek, Anthony. *The Walking Liberty Half Dollar,* Sanford J. Durst Publishing, Long Island City, NY, 1983

Yeoman, R.S. *A Guide Book of United States Coins* - 54th Edition, Whitman, St. Martin's Press, NY, NY, 2001

II. American History

Beliles, Mark A. & Stephen K. McDowell. *America's Providential History*, Providence Foundation, Charlottesville, VA, 1989

Duncan, Dayton & Ken Burns. *Lewis & Clark - An Illustrated History*, Random House, Inc. New York, 1997

Foster, Marshall & Mary-Elaine Swanson. *The American Covenant,* Mayflower Institute, Roseburg, OR, 1981

Kirshon, John. *Chronicle of America*, Prentice Hall Trade, NY, NY, 1990

Marshall, Peter. *The Light & the Glory*, Zondervan Publishing, Grand Rapids, MI, 1988

Ryter, Jon C. *What Ever Happened to America?*, Hallberg Publishing, Tampa, FL, 2000

III. Economic/Monetary History

Jacobson, Steven. *Wake Up America*, MCIA Media, Winston-Salem, NC, 1995

Kershaw, Peter. *Economic Solutions*, Quality Press, Boulder, CO, 1997

McMaster Jr., R.E. *No Time For Slaves*, Reaper Publishing, Phoenix, AZ, 1986

Mullins, Eustice. *The Secrets of the Federal Reserve*, Bankers Research Institute, 1991

Peacocke, Dennis. *Doing Business God's Way*, SCS Publishing, Santa Rosa, CA 1993

Pick, Franz. *An Advance Obituary, The U.S. Dollar,* Pick Publishing, NY, NY, 1981

Reisman, George. *Capitalism, A Treatise on Economics,* Jameson Books, Ottawa, IL 1990

Skousen, Mark. *Economics on Trial,* Phillips Publishing, Potomic, MD, 1992

IV. U.S. Coin Market Periodicals

Certified Coin Dealer Newsletter, CDN Publishing, Torrance, CA (weekly - $117/yr.)

Coinage, Miller Magazines, Ventura, CA (monthly - $24/yr.)

Coin Dealer Newsletter, CDN Publishing, Torrance, CA (weekly - $117/yr.)

Coin World, Amos Press, Sidney, OH (weekly - $34.95/yr.)

The Numismatist, American Numismatic Association (ANA) Colorado Springs, CO (monthly - $39/yr.)

V. Swiss America Resources
Print
A Rare Opportunity, brochure, 1990-2001

Real Money Perspectives, newsletter, 1987-2001
Rare Coin Buyer's Guide, brochure, 1996-2001
U.S. Gold Commemorative Research Report, white paper, 2000-01
Proof Silver Type (1936-1942) Research Report, white paper, 2001
The Right to Own Gold, white paper, 1995-2001
True Wealth, by Craig R. Smith, white paper, 1996-2001
Guide to Investing in Gold, brochure, World Gold Council, 1998-2001

Multimedia

Investing Wisely in the Next Millennium, CD, cassette, 1999-2001
CNNfn Interviews Craig R. Smith, video & CD, 2000-2001
The Big Picture, audio CD & booklet, 2000-2001
Rediscovering Gold with Pat Boone, CD, cassette, 2001

VI. Online Resources

Gold - U.S. Rare Coins

www.buycoin.com
www.swissamerica.com
www.gata.org
www.wgc.org
www.greysheet.com
www.pcgs.com
www.ngccoin.com

Economics/Market News

www.true-wealth.com
www.buycoin.com
www.cnnfn.com
www.newsmax.com
www.dailyreckoning.com
www.dowtheoryletters.com
www.gostrategic.org
www.mises.org